The Mechanisms of
Job Stress
and Strain

The Mechanisms of
Job Stress
and Strain

John R. P. French, Jr.
Robert D. Caplan
R. Van Harrison

The University of Michigan

1807 1982

JOHN WILEY & SONS

Chichester · New York · Brisbane · Toronto · Singapore

Library of Congress Cataloging in Publication Data:

French, Jr., John R. P.
  The mechanisms of job stress and strain.

  (Wiley series on studies on occupational stress)
  Includes index.
  1. Job stress.   2. Stress (Psychology)   I. Caplin,
  Robert D.   II. Van Harrison, R.   III. Title.
  IV. Series.
  HF5548.85.F73   .58.7   81-21871
                                        AACR2
  ISBN 0 471 10177 X

British Library Cataloguing in Publication Data:

French, Jr., John R. P.
  The mechanisms of job stress and strain. —
  (Wiley series on studies in occupational stress)
  1. Job stress 2. Stress (Psychology)
  3. Stress (Physiological)
  I. Title   II. Caplan, Robert, D.   III.
  Harrison, R. Van
  158.7   HF5548.8

  ISBN 0 471 10177 X

Phototypeset by Dobbie Typesetting Service, Plymouth, Devon, England
and Printed in the United States of America.

Acknowledgements

Considerable thanks are due to the officials of labor and management who gave approval to this study, and to the many members of the work force who gave us their time. They and their organizations remain anonymous in accordance with the conditions of this study.

This book represents an extension of hypothesis testing and analyses of data collected in a prior study, *Job Demands and Worker Health*. That study involved the contributions of several other persons in addition to the authors of this book. Sidney Cobb, MD, and Richard Pinneau, PhD, were two of these people. Drs Cobb and French developed the initial ideas for the study in collaboration with Drs William Kroes and Bruce Margolis of the National Institute for Occupational Safety and Health (NIOSH). As the study progressed, Sidney Cobb provided the expertise in theory and method required to investigate issues of psychosomatic epidemiology. Pinneau and Harrison shared responsibility, along with others, for the design of the questionnaire, securing of research sites, data management, and analyses. French's presentation of theory (Chapter I) and future directions (Chapter VII) represent his commitment to a programmatic approach to the study of the social environment and its effects on human wellbeing. Harrison focused on the development and testing of theory regarding person–environment fit (Chapter III). Caplan coordinated the study, and wrote the chapters covering methodology and the multivariate analyses of main and interaction effects and the interpretations of those effects (Chapters II and IV–VI).

The survey data were collected with the help of the Survey Research Center Field Staff. Melinda Wagner performed much of the data management, coding, and preparation of tables. Terry Roth was responsible for collecting and for analyzing the blood samples in the laboratory. Joseph Hurrell of NIOSH arranged for the collection of data from police. The staff from the Institute for Social Research arranged and managed all other data collection.

The initial survey was conducted under a contract (HSM-99-72-61) and a grant (1 R01 OH 00563-01) from the National Institute for Occupational Safety and Health (NIOSH), US Department of Health, Education, and Welfare. Further analyses and writing of this book were made possible by University of Michigan Sponsored Computer Time to the authors and by a National Institute of Mental Health Career Award (K06 MH 21844) to John R. P. French, Jr. Diane Palmer typed the original manuscript from which this book was set.

Contents

Editorial Foreword to the Series

This book, *The Mechanisms of Job Stress and Strain*, is the seventh book in the series of *Studies in Occupational Stress*. The main objective of this series of books is to bring together the leading international psychologists and occupational health researchers to report on their work on various aspects of occupational stress and health. The series will include a number of books on original research and theory in each of the areas described in the initial volume, such as Blue Collar Stressors, The Interface Between the Work Environment and the Family, Individual Differences in Stress Reactions, The Person–Environment Fit Model, Behavioural Modification and Stress Reduction, Stress and the Socio-technical Environment. The Stressful Effects of Retirement and Unemployment and many other topics of interest in understanding stress in the workplace.

We hope these books will appeal to a broad spectrum of readers — to academic researchers and postgraduate students in applied and occupational psychology and sociology, occupational medicine, management, personnel, etc. — and to practitioners working in industry, the occupational medical field, mental health specialists, social workers, personnel officers, and others interested in the health of the individual worker.

CARY L. COOPER,
*University of Manchester Institute of
Science and Technology (UK)*

STANISLAV V. KASL,
Yale University

Chapter I

The Theoretical Model

Introduction

In all types of jobs there is an interplay between the demands that the job makes on the employee and the demands that the employee requires of the job. The lack of accommodation between the demands of employees and those of their organizations is both a significant social issue and an important topic for research.

Research on the accommodation of organizational members and organizations to one another can focus on outcomes such as productivity, profits, employee commitment to the organization, and the accomplishment of organizational goals. These are obviously important outcomes for the well-being of both the organizational member and the organization. In this book we report on a program of research that has examined another important outcome of the interaction between work organizations and their members — the mental and physical health of the employees. Rather than studying physical hazards on the job, we have examined psychosocial aspects of organizations which may influence employee health. To study psychosocial factors at work and the link between them and the employee, we have used the concept of goodness of fit between the person's characteristics and the characteristics of the organization. More will be said about this concept shortly.

Clearly the attainment of the most healthful work environments requires an understanding of how one can reduce the effects of physical and chemical toxins as well as psychosocial stressors. Satisfaction with relationships and activities in the work setting are a bitter victory if they are cut short by disabilities due to physical hazards. On the other hand, a chemically pristine environment is only minimally rewarding if employees must constantly fight back the pangs of boredom or the stress of being pushed beyond their intellectual and motivational limits.

This book is the second of two volumes reporting a large study of occupational stress, strain and mental health. The first volume (Caplan *et al.*, 1980) described occupational differences in stress and strain among 23 occupations. It also reported the correlations between environmental stresses and strains within the person in this diverse sample of 2010 men. This second

1

volume concentrates on multivariate analyses of the effects of stress on strain. It also attempts to test more definitively a general theoretical model which, with increasing specification, has guided our program of research during the past twenty years.

Our research program on the Social Environment and Mental Health aims to discover those environmental variables which have a significant effect on health and which could be altered so as to improve the quality of life. Thus, there is an emphasis on discovering environmental factors that prevent both maladjustment and excessive strain in normal people. Following Lewin (1951), we assume that it is the interaction of environmental variables with relevant properties of the person that determines behavior, strain and mental health.

Purposes of the Theoretical Model

The model used in this study incorporates four general aims that are part of our research program on the Social Environment and Mental Health.

1. It attempts to *quantify* both occupational stress and mental health along a variety of dimensions on the grounds that both stress and health are *multidimensional.*
2. It focusses on the range of mental health attributes found in the *normal* population. As a result it considers some of the positive conceptions of mental health (Jahoda, 1958) such as satisfaction with work as well as negative states such as anxiety, depression, and somatic complaints.
3. It presents a *systematic theoretical framework* for quantifying the inter-relationships among four dimensions of mental health including the distortions in perceiving one's work environment and one's own abilities and motives. As will be explained below, the model has the potential for quantifying the adjustive process in terms of coping and defense.
4. The framework provides a *guide for generating hypotheses* about the relationships among the job environment, the nature of the employee, and resultant health consequences.

The model presented in Figure I.1 is an elaboration of one described by French, Rodgers and Cobb (1974). It maps many more relationships than can be examined in the present cross-sectional study and relies primarily on self-reported data. For example, the model presents hypothesized relationships involving objective characteristics of the work environment and objective measures of the person including motives, abilities, defenses, and coping efforts. No direct, objective measures are available for these parameters in this study. These unmeasured parts of the model are introduced because we need to be aware of their roles when we interpret the results concerning the measured parts of the model. Additionally, the model in Figure I.1 serves as a framework for organizing the presentation of the hypotheses which are tested. The reader should find it useful to refer to Figure I.1 as a guide to all that follows.

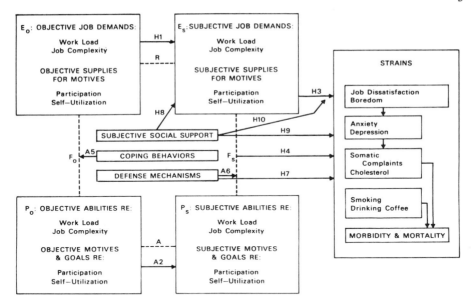

Figure I.1   A model for the effects of stressful job environments on subjective stresses, on strains, and on morbidity and mortality. Each box represents a different type of variable (described in capital letters) but only a few exemplary variables are presented (in small type). Arrows represent direct causal relations; an arrow terminating on a second arrow represents a conditioning effect. Broken lines represent the discrepancy between variables in the connected boxes (see text); they denote four dimensions of mental health

The Basic Elements of the Model

The four square boxes in Figure I.1 contain the four basic elements from which we construct two types of person–environment fit and two types of accuracy of perception. We start with the metatheoretical distinction between objective variables (the two square boxes on the left) and the subjective variables (the two square boxes on the right). Objective variables exist for the scientist and are based on scientific observations. These observations may include quite complex and indirect methods of observation such as various forms of intelligence tests or physiological measurements such as the electrocardiogram. Subjective variables, on the other hand, exist for the person as self-perceptions and perceptions of the objective world with which the person is in contact. They may be accurate or distorted pictures of this world, but in either case we assume that the person's responses are directly determined by these subjective variables.

Both objective and subjective variables can refer either to the environment, E (the top two square boxes in Figure I.1), or to the person, P (the corresponding bottom two boxes). Within the objective environment, E_o, we distinguish two classes of variables: (1) environmental *demands* which initiate action by the person (such as performance of work tasks), and (2) environmental *supplies* for the person's motives (such as income). Similarly, the subjective environment,

E_s, contains both perceived job demands and perceived supplies for motives.

The objective person, P_o, contains two classes of variables which correspond to the two classes of environmental variables (1) objective *abilities* to meet the demands of the environment (such as the knowledge and skill to handle the workload), and (2) *motives* in the person which initiate action by the person to obtain supplies from the environment—these include needs, values, goals and preferences. For example, the need for self-utilization can motivate an individual to seek out opportunities to employ valued abilities. The subjective person, P_s, contains perceived abilities and motives that correspond, more or less veridically, to the objective ones.

The Four Dimensions of Mental Health

We can generate from the above basic elements four difference scores (see the four broken lines in Figure I.1) that represent important dimensions of mental health. This is possible if and only if the four basic elements can be conceptualized and quantified on *commensurate dimensions*; that is, they must be measurable along the same scales. We can illustrate this principle of commensurate dimensions with the case of a typist. We shall consider only one dimension, typing speed, and the demands and abilities related to it. However, we should keep in mind that a similar case could be presented to illustrate commensurate dimensions of needs and supplies for these needs.

Let us suppose that the objective workload in the typist's job demands a typing speed of 70 words per minute, but the corresponding objective ability of this particular typist is only 60 words per minute. Then the *objective fit* (F_o in Figure I.1) of this typist can be quantified as a deficiency of exactly ten words per minute. The typist might adjust to this misfit either by defensively distorting upward the subjective ability to match the required 70 words per minute or, alternatively, by distorting downward the objective job demand so that the subjective demand is perceived to be only a speed of 60 words per minute. Either of these defensive maneuvers, or some combination of the two, could result in a perfect *subjective fit* (F_s) where the perceived demands exactly matches the perceived abilities. However, the distortion of the typist's ability simultaneously reduces *accessibility of the self* (A in Figure I.1), another important dimension of mental health. The distortion of the environmental demand likewise reduces *contact with reality* (R in Figure I.1). Thus the four basic elements of objective and subjective demands and abilities, when measured commensurately in terms of words per minute, generate four dimensions of mental health: (1) objective person–environment fit (F_o); (2) subjective person–environment fit (F_s); (3) accessibility of the self (A); and (4) contact with reality (R). In this model these four dimensions of mental health form a dynamic system in the sense that a change in any one of them will necessarily produce a change in at least one other. In the case described above, for example, a change in perceived ability (P_s) produces both an improved subjective fit and a worsened accessibility of the self; also a decreased subjective demand for typing speed results in both an improved subjective fit and a worsened contact with reality. In both instances, the magnitude and the direction of these interdependent changes can be deduced from the model (French, 1978).

Now we are in a position to specify our usage of the terms 'stress' and 'strain'. 'Stress' is not a technical concept in our theory; instead it is a commonly used word referring to any of the following technical concepts: (1) objective misfit (F_o); (2) subjective misfit (F_s); (3) a variable in the objective environment (E_o) which is presumed to pose a threat to the person; and (4) a variable in the subjective environment (E_s) which the person perceives as threatening. *Chronic* stress refers to a relatively stable state of the environment or of person–environment misfit; *acute* stress refers to a sudden increase in stress. The study reported here deals primarily with chronic job stresses. The reader should keep in mind the possibility that acute stresses may have somewhat different effects. *Strain* is defined as any deviations from the normal state or responses of the person. These include psychological strains such as job dissatisfaction and anxiety, physiological strains such as high blood pressure, and behavioral symptoms of strain such as excessive smoking and consumption of alcohol. Continued high levels of these strains can eventually affect the levels of morbidity and mortality.

Assumptions and Hypotheses

All forms of stress are hypothesized to produce some form of strain. However, no particular stress is expected to affect *all* forms of strain; instead we generally expect a certain degree of specificity in the effects of stresses on strains. For example, we expect that a heavy workload will affect dissatisfaction with workload more strongly than it will affect general job dissatisfaction. The general hypothesis states that a stress will have stronger effects on *more relevant* strains. These more specific hypotheses and findings are presented in later chapters.

The more general assumptions and hypotheses of the model are presented in Figure I.1 as arrows running from one box to another. The assumptions represent causal relationships which are not tested in this study, whereas the hypotheses represent causal relations which *are* tested in this study. In both cases, these propositions deal with *direct* causal relations where no intervening variables are assumed. It should be noted that each arrow represents a large number of propositions because each box in the diagram represents a whole class of specific variables. Thus, only a few of the many specific hypotheses in this study are illustrated in Figure I.1.

The general psychosocial theory of strain and health modelled in Figure I.1 asserts that the interaction of objective variables in the environment and objective variables in the person affect corresponding interactions between subjective properties of the environment and subjective properties of the person to influence strain and health. The more specific assumptions and hypotheses, numbered to correspond to Figure I.1 are as follows:

6

Hypothesis 1 (see arrow H1)   Objective job demands and supplies for one's needs will tend to produce corresponding perceived demands and supplies.

Assumption 2 (see arrow A2)   Objective abilities and goals will tent to produce corresponding subjective abilities and goals.

Hypothesis 3 (see arrow H3)   High demands or low supplies in the subjective environment (E_s) will cause psychological strains such as job dissatisfaction and anxiety, behavioral strains such as smoking, and physiological strains such as high blood pressure.

Hypothesis 4 (see arrow H4)   The greater the subjective misfit (F_s) between the person's subjective abilities and goals and the corresponding subjective job demands and supplies, the greater the psychological, physiological, and behavioral strain. Misfit includes both *too much* and *too little* ability or supplies. The form of this relation is discussed in greater detail in Chapter III.

Assumption 5   Various coping activities will reduce objective misfit (F_o) between the person and the environment.

Assumption 6   Defense mechanisms will tend to reduce subjective misfit (F_o) by distorting the perception of the environmental stress and/or of the self without any corresponding changes in objective misfit (F_o). This is in opposition to the tendencies toward accurate perception of the environment and of the self, which are stated in Hypothesis 1 and Assumption 2 above.

Hypothesis 7   Defense mechanisms will reduce strain.

Hypothesis 8   Subjective social support will reduce stresses in the subjective environment.

Hypothesis 9   The greater the subjective social support, the lower the strain.

Assumption 10   To the extent that there is high subjective social support, subjective stresses will not produce the usual strain, i.e. social support will moderate the effects of Hypothesis 3.

Hypothesis 11   There are various direct and indirect causal relations among strains (see the arrows within the box on strains in Figure I.1). These hypotheses are discussed in more detail in Chapter VI.

Causal Pathways and Derived Hypotheses

In Figure I.1 and in the above propositions there are many causal pathways by which an independent variable affects an intervening variable which in turn affects a dependent variable. In some of these paths there are several intervening variables. In all such causal sequences of unidimensional variables we can logically deduce that there must be some indirect causal connections between two variables even when there is no direct causal relation (and therefore no arrow in Figure I.1). If A causes B and B causes C, then it is necessarily true that A exerts an indirect causal influence on C. Such derived hypotheses, implicit in Figure I.1, can explain some of the well-known *occupational differences* in strain, in morbidity, and in mortality (Pflanz, Rosenstein, and Von Uexkul, 1956; Russek, 1965).

The logic of the *derived* explanations implicit in Figure I.1 can be illustrated as follows. There are objective differences among jobs. The differences occur for such variables as the amount of workload or the complexity of the work. These differences cause occupational differences in corresponding subjective job stresses which in turn cause occupational differences in the job-related strains such as job dissatisfaction. In turn, the job-related strains cause general affects such as anxiety and depression. General affective states then cause physiological and behavioral strains which eventuate in morbidity and mortality. Therefore, different occupations will have different patterns of strain and of health.

Our own studies show how this theory of derived relationships helps to explain occupational differences in strain and health (see Chapter VI and Caplan, 1972; Cobb and Rose, 1973; House, 1972; Sales and House, 1971; Quinn *et al.*, 1971). The theory can also help to account for social class gradients in mental illness and psychosomatic diseases (Kasl and French, 1962) because occupation is an important component of social class and socioeconomic status.

Reciprocal Causation and Causal Cycles

Although the direction of causation illustrated in Figure I.1 is generally one way, we believe this is too simple. The causal relations may also be reciprocal. For example, we have found a correlation between subjective social support and depression (Caplan *et al.*, 1980) and we have interpreted this finding to reflect both the tendency for low social support to increase depression and the tendency for depression to produce an underestimation of social support. The two variables form a causal cycle or perhaps even a vicious spiral: a loss of social support causes depression, which lowers still further the subjective support, which in turn causes still further depression and so on. In cross-sectional studies such cycles cannot be evaluated adequately. In this study, we must be especially cautious about asserting that occupational variables change P_o and P_s because it is likely that some job characteristics may cause the person to select a particular occupation or to be selected for a particular job.

We note that there are many causal cycles involving more than two variables which are included in our model but which, for the sake of clarity, have not been explicitly shown in Figure I.1. For example, we have shown that stress causes strain, but we have not indicated that strain initiates coping behaviors and defense mechanisms and that these in turn affect both the objective stresses and the subjective stresses thus completing a causal cycle. The cross-sectional design of the present study cannot disentangle such causal cycles. However, they must be considered when interpreting our associational results.

Plan of the Book

The remaining chapters of this book are primarily devoted to testing different parts of the theoretical model. After a chapter on methods, we present in Chapter III an elaboration and tests of Hypothesis 4 on the effects of person-environment fit on strains. Chapter IV deals with derived hypotheses and causal chains linking job stresses to job-related strains which in turn affect general strains (such as anxiety) and somatic complaints (see Hypothesis 10). In Chapter V we examine four hypotheses (not shown in Figure I.1) about variables which might condition the effects of stress on strain. We return in Chapter VI to the initial starting point of our model: the objective job demands and supplies which, according to Hypothesis 1, produce corresponding subjective demands and supplies. Here we test the adequacy of our theory and measures of E_s, P_s, and F_s to explain the obtained occupational differences in strains. Finally, Chapter VII summarizes the empirical results from previous chapters, points out key implications for further research, and discusses in more detail the implications of our findings and theory for dealing with problems of stress and strain in work organizations.

Chapter II

Review of the Methodology

The methods and sample characteristics of the study are described in detail in Caplan *et al.* (1980). Only a brief review of the methodology is presented here.

Sample

Sampling Occupations

Table II.1 presents a list of the 23 occupations studied in this project and their sample sizes. The occupations were chosen to represent a wide array of job demands, or were selected because previous studies indicated that the occupations had high levels of psychosomatic strain. For example, air traffic controllers were chosen for study because of their high rates of peptic ulcer and high blood pressure (Cobb and Rose, 1973). Some occupations were chosen because they had low stress. Consequently, the occupations studied do not represent a random sample of occupations in the American workforce, but they do represent a wide range of types and amounts of stress.

A comparison of occupational differences in measures of demographic characteristics, job demands, personality traits, and psychological and physiological strain indicates that we were successful in sampling a broad range of job demands and worker health (Caplan *et al.*, 1980).

Occupation categories. Once the sample was obtained, occupation was determined by coding three pieces of self-reported information: (a) main occupation, (b) section of the organization where the person worked, and (c) a brief description of what the person did as part of the job.

Physical hazards avoided. Occupations chosen for the study were limited to those where exposure to physical and chemical hazards were minimal so that we could have a clearer focus on the effects of social and psychological hazards of work. For example, in the sample of forklift drivers, operators of electric and propane vehicles were included but operators of gasoline-powered vehicles were excluded because of the exposure to carbon monoxide. Checks with union and management officials at sites indicated that carbon monoxide and noise levels were within OSHA standards. Our study did not have budgetary provisions for independent verification of this information.

Table II.1   Occupational groups and their sample sizes

Occupational groups	Abbreviations used in tables	N
Blue-collar[a]		
Forklift driver	Forklift drvr	46
Assembler, machine-paced	Assemb mach	79
Assembler, machine-paced relief	Assemb relief	27
Assembler, nonmachine-paced	Assemb nomach	69
Machine tender	Mach tender	34
Continuous flow monitor	Contin flow	101
Delivery service courier	Courier	20
Tool and die maker	Tool and die	77
Blue/white-collar		
Electronic technician	Elec tech	93
Policeman	Policeman	111
Train dispatcher	Dispatcher	86
Blue-collar supervisor	Sup blue coll	178
White-collar supervisor	Sup whte coll	42
White-collar		
Air traffic controller, large airports	ATC, large	82
Air traffic controller, small airports	ATC, small	43
Programmer	Programmer	90
Accountant	Accountant	92
Engineer	Engineer	110
Scientist	Scientist	117
Professor	Professor	74
Administrative professor	Admin prof	25
Administrator	Administrator	253
Family physician	Physician	104
Miscellaneous, gathered incidentally	Miscellaneous	57
Total		2010

[a]This ordering of occupational groups reflects — with some minor changes — an arrangement from lowest to highest Duncan SES score and therefore from lowest to highest socioeconomic status.

Organizational sites. The 23 occupations were obtained from 67 different sites. At least two organizational sites were obtained for almost every occupation so that occupation and work organization would not be confounded. The train dispatchers were drawn from a wide variety of track conditions in the southern, midwestern and eastern United States.

Characteristics of the Samples of Respondents

All respondents were male, primarily white, and had worked in their current job for at least six months. All respondents volunteered and formed a sample of opportunity. Three types of samples were used in the analyses. They are as follows.

Total sample. This sample consisted of all 2010 respondents. It was used mainly where the means of the occupations were used as raw data for analyses. When analyses were confined within any one occupation, we used the total set of respondents for that occupation. In the latter case, all the persons samples in that occupation were studied.

Physiological sample. A subset of the total sample, 390 persons, entered this sample. Table II.2 describes the eight occupational groups included in that sample. These groups were chosen either because they represented different degrees and types of stress or because they had high rates of psychosomatic disease. For example, machine-paced assemblers were selected because of the

Table II.2   Occupations with physiological data

Occupational group	Sample N	Double sample N^a
Assembler, machine-paced	46	28
Electronic technician	64	16
Supervisor/blue- and white-collar	25	15
Air traffic controller, large airport	82	33
Air traffic controller, small airport	41	23
Scientist	55	—
Administrator	62	—
Miscellaneous	15	—
Total	390	115

[a]The number of persons from the left column who had a second blood sample drawn 20 minutes after the first.

research literature (see, for example, Blauner, 1964) which suggests that among blue-collar employees these assemblers have high job stress. Administrators and scientists were chosen because of previous research (French and Caplan, 1970) suggesting that they represented high and low rates respectively of coronary heart disease.

Random stratified sample. This sample was an occupationally stratified random subsample of the 2010 men. The total sample did not represent all occupations equally because some occupations had only small numbers of persons in the sample and other occupations had larger numbers of persons. For example, there were 253 administrators but only 20 delivery service couriers. To maximize the diversity of occupational characteristics, we randomly selected approximately fourteen persons from each occupational group. The resulting subsample consisted of 318 persons.

Characteristics of the Samples

For the characteristics that follow there was little difference between the total sample and the random sample in practically all cases.

The random sample had a mean age of 39, averaged a high school education, had been on the job on average at least one year, and had worked 45 hours per week on the average. Mean gross earnings for the sample were $17,379 in 1972, the year prior to the data collection.

Compared to an occupationally stratified, random sample of the United States (Quinn *et al.*, 1971) our random stratified sample differed in several ways. Our sample was almost all white, male, more highly paid, more highly educated and tended to underrepresent the oldest and youngest members of the workforce of the United States.

The members of the physiological sample, compared to the random sample, tended to be somewhat younger, somewhat more educated and more highly paid (reflecting the overrepresentation of professionals in the physiological sample). The employees in the physiological sample worked somewhat fewer hours (42 hours per week), reflecting the large number of air traffic controllers in the sample whose maximum number of hours was limited by FAA regulations. Length of service on the job was very similar to the random sample.

Method of Data Collection

Administering the Questionnaire

Respondents were informed of the study primarily at their places of work. Contact was usually by a letter indicating management and union endorsement of the study. The letter and the cover of the questionnaire informed the employee that all responses would be anonymous and confidential.

The questionnaire could be completed in about one hour by a person with a high school education. Persons who read a lot as part of their work (such as air traffic controllers and administrators) completed the questionnaire in 30 to 40 minutes. In some instances the questionnaires were completed at the sites of work. In other instances they were completed at union halls or at home and mailed in.

It was not possible to determine the response rates exactly for each occupational group. Caplan *et al.* (1980) describes those occupational groups for whom the percentage of persons completing the questionnaires appeared to be excellent and occupations for whom the percentage was lower. Several occupations, such as train dispatchers, had response rates close to 100 per cent. In several other occupations, such as policemen and university professors, one-half to three-quarters of those notified subsequently responded. In blue-collar occupations in particular, as few as 25 per cent of those employees notified filled out questionnaires. Accordingly, one should be cautious in generalizing from data on any specific occupation to all persons in that occupation.

Collecting the Physiological Data

Each employee in the physiological sample first completed the questionnaire. Next, the employee was interviewed privately to determine the consumption of food, medication, caffeine, and tobacco during the preceding 24 hours. Then blood pressure was measured. Finally, 15 cc of blood were obtained. When possible a second blood sample was taken on every other respondent to examine test–retest reliability of the determinations. The blood samples were allowed to clot at room temperature, were then refrigerated, and a few hours later were spun to separate out the serum.

Measures

Table II.3 lists all the measures used in the study except the physiological measures. The table also gives data on the cross-sectional estimates of reliability of multi-item indices, the sources of the indices if they were used in previous research, and the abbreviations that are used to refer to the measures when they are presented in tables and figures. Table II.4 lists the physiological measures and information on their reliabilities. Caplan *et al.* (1980, Appendix H) present the intercorrelations among almost all the measures in the study. Appendix E of Caplan *et al.* (1980) gives the item content and scoring of all the questionnaire measures except for the content of five items included in Appendix B of this volume (page 129). Those five items are noted later in this chapter.

Construction of Indices

Several criteria were used to construct multi-item indices. Each item had to correlate significantly with other items measuring the same construct, and this correlation had to be higher on the average than its correlation with items measuring other constructs. Most items within an index had to show similar interrelations within a blue-collar occupation (machine-paced assembler) and within a white-collar occupation (professor). This last criterion assured high index reliability even when the index was used to assess the nature of the work in markedly different work environments. Of 152 items constructed for such indices, only twelve items failed to meet these criteria, and these were excluded from further study.

Whenever the text refers to a specific measure or index, the first letters of the index are capitalized. Furthermore, the index is named for the end of the scale with the most stress or strain, such as 'Job Dissatisfaction' rather than 'Job Satisfaction'.

Measures of Subjective Environment

The estimated reliabilities of these measures of stress ranged from .71 to .89. The indices are briefly described below with sample items presented parenthetically. Most indices required ratings on 5-point scales with dimensions ranging from 1 = 'very little' to 5 = 'a great deal' or from 1 = 'not at all' to 5 = 'very much'.

Table II.3 Measures of demographic characteristics, subjective environment (stress), personality, P–E fit, psychological strain, health-related behavior and illnesses

Measures	Abbreviations used in tables	Number of items	Cross-sectional estimate of reliability[a]	Source
Demographic characteristics				
Occupation	Occupation	3	—	
Age	Age	1	—	
Years of Education-P	Education-P	1	—	(Labelled 'Schooling' in Caplan et al., 1980)
Duncan Socioeconomic Status	Duncan SES	1	—	Scheffler, Rice and Kaplan (1971); Reiss et al. (1961)
Income-E	Income-E	1	—	
Length of Service-P	Service-P	1	—	(Labelled 'Length of Service' in Caplan et al., 1980)
Subjective environment (Stress)				
Typical Education Needed-E[b]	Education-E	1	—	
Typical Length of Service Needed-E[b]	Service-E	1	—	
Number of Persons Supervised[b]	Number Supervises	1	—	ISR National Surveys
Hours Worked per Week	Hrs Wkd/Week	1	—	ISR National Surveys
Hours of Overtime per Week	Overtime-E	1	—	
Quantitative Workload-E	Qnt Wk Ld-E	7	.71	Caplan (1971); Quinn et al. (1971)
Workload, Quinn	Work Ld-Quinn	4	.85	Additional workload items from Quinn and Shepard (1974)
Variance in Workload	Variance Wk Ld	3	.79	Based on presurvey interviewing
Responsibility for Persons-E	Resp Person-E	4	.89	Caplan (1972), presurvey interviewing
Job Complexity-E	Complexity-E	6	.72	Theoretical and empirical research of Hackman and Lawler (1971); Kohn (1969); Quinn and Shepard (1974)
Concentration	Concentration	1	—	
Role Conflict	Role Conflict	3	.80	Theoretical and empirical research of Kahn et al. (1964); Kahn and Quinn (1970)
Role Ambiguity-E	Role Ambig-E	4	.84	Kahn et al. (1964); Caplan (1972)
Job Future Ambiguity	Future Ambig	4	.79	Vickers (1979); Quinn et al. (1971)
Job Insecurity	Job Insecurity	1	—	

Measure	Abbreviation	N	Reliability	Source
Underutilization of Abilities	Underutilzat	3	.85	Taylor and Bowers (1972); Caplan (1971)
Inequity of Utilization[b]	Ineq Utilzat	1	—	
Inequity of Pay	Ineqty of Pay	3	.81	Theoretical and empirical research of Adams (1965); Quinn et al. (1974)
Participation	Participation	3	.80	Likert (1961); Caplan (1972); and modifications based on theoretical conceptualizations
Social Support from Supervisor	Support Sup	4	.83	The support measures are based on theoretical and empirical research of Pinneau (1972); Taylor and Bowers (1972); Likert (1961); Gore (1974)
Social Support from Others at Work	Support Othrs	4	.73	
Social Support from Wife, Friends and Relatives	Support Home	4	.81	
Personality				
Sales Type A Personality	Sales Type A	9	.74	Sales (1969); Vickers (1973)
Flexibility	Flexibility	7	.71	Gough (1957)
Assert Good Self	Assert Good	7	.57	Crowne and Marlowe (1964); Lillibridge (1970)
Deny Bad Self	Deny Bad	7	.62	Crowne and Marlowe (1964); Lillibridge (1970)
Personal preferences				
Quantitative Workload-P	Qnt Wk Ld-P	7	.60	The general theoretical formulation of the 'P' and 'Fit' indices were developed from French, Rodgers and Cobb (1974). The source of the content for each specific measure is listed after the 'E' form of the measure (e.g., for Quantitative Workload-P, see Quantitative Workload-E). See also Appendix D in Caplan et al., 1980
Responsibility for Persons-P	Resp Persons-P	4	.87	
Job Complexity-P	Complexity-P	6	.71	
Role Ambiguity-P	Role-Ambig-P	4	.86	
Income-P	Income-P	1	—	
Overtime-P	Overtime-P	1	—	
Person-Environment-Fit				
Job Complexity-Fit	Complexity-Fit	6	.54	The general theoretical formulation of the 'P' and 'Fit' indices were developed from French,
Job Complexity-Deficiency	Complexity-Defic.	6	.84	

Table II.3 (continued)

Measures	Abbreviations used in tables	Number of items	Cross-sectional estimate of reliability[a]	Source
Job Complexity Excess	Complexity Excess	6	.76	Rodgers and Cobb (1974). The source of the
Job Complexity Poor Fit	Complexity Poor Fit	6	.72	content for each specific measure is listed after
Role Ambiguity Fit	Ambiguity Fit	4	.82	the 'E' form of the measure (e.g. for
Role Ambiguity Deficiency	Ambiguity Defic.	4	.87	Quantitative Workload-P, see Quantitative
Role Ambiguity Excess	Ambiguity Excess	4	.82	Workload-E). See also Appendix D in Caplan
Role Ambiguit Poor Fit	Ambiguity Poor Fit	4	.74	et al. (1980).
Responsibility for Persons Fit	Resp Person-Fit	4	.83	
Responsibility for Persons Deficiency	Resp Person Defic.	4	.88	
Responsibility for Persons Excess	Resp Person Excess	4	.77	
Responsibility for Persons Poor Fit	Resp Persons Poor Fit	4	.74	
Workload Fit	Workload Fit	7	.70	
Workload Deficiency	Work Load Defic.	7	.77	
Workload Excess	Work Load Excess	7	.88	
Workload Poor Fit	Work Load Poor Fit	7	.54	
Income Fit	Income Fit	1	—	
Overtime Fit	Overtime Fit	1	—	
Service Fit	Service Fit	1	—	
Service Deficiency	Service Defic.	1	—	
Service Excess	Service Excess	1	—	
Service Poor Fit	Service Poor Fit	1	—	
Education Fit	Education Fit	1	—	
Education Deficiency	Education Deficiency	1	—	
Education Excess	Education Excess	1	—	
Education Poor Fit	Education Poor Fit	1	—	
Psychological strain				
Job Dissatisfaction	Job Dissat	4	.85	Quinn and Shepard (1974)
Workload Dissatisfaction	Wk Ld Dissat	3	.82	Theoretically derived for this study
Boredom	Boredom	3	.86	Theoretically derived for this study

Somatic Complaints	Somat Complnts	10	.76	Gurin, Veroff and Feld (1960), Langner (1962)
Depression	Depression	6	.83	Cobb (1970); Zung (1965)
Anxiety	Anxiety	4	.75	Cobb (1970). The items also overlap with those in Gurin, Veroff and Feld (1960) and Spielberger, Gorsuch and Lushene (1970)
Irritation	Irritation	3	.80	Cobb (1970)
Health-related behavior				
Smoker–Nonsmoker	Smoker	1	—	
Ex-Smoker	Ex-Smoker	1	—	
Number of Cigarettes Smoked (if > 0)	Cig Smoked > 0	1	—	
Number of Cigarettes Smoked in Last 4 Hours	Cig 4 Hours	1	—	Physiological sample only
Cups of Coffee	Coffee	2	—	
Caffeinated Drinks	Caffein Drnks	3	—	
Caffeinated Drinks in Last 4 Hours	Caff 4 Hours	1	—	Physiological sample only
Obesity Index	Obesity	2	—	
Recency of Dispensary Visit	Disp Visit	1	—	
Staffed Dispensary Visit	Stffd Dis Vis	1	—	Only for persons indicating they have a staffed dispensary
Illnesses				
Cardiovascular Disease	Cardiovascular	2	—	
Peptic Ulcer	Peptic Ulcer	2	—	
Gastrointestinal Problems	Gastrointest	2	—	
Respiratory Infection	Respiratory	2	—	

[a]The estimate of reliability is computed using the following formula (cited in Nunnally, 1967):

$$r_{kk} = \frac{k\bar{r}_{ij}}{1 + (k-1)\bar{r}_{ij}}$$

where k is the number of items in the test, and $\bar{r}_{ij}$ is the average intercorrelation between items in the test. This is based on the occupationally stratified random subsample of the total sample, $n = 318$. The item content of all other measures is presented in Appendix B, page 132.

[b]The item content for this measure is presented in Appendix E of Caplan et al. (1980).

Quantitative Workload-E (E for subjective environment) refers to the *amount* of work the person is given ('How much time do you have to do all your work?') in contrast to the *qualitative* difficulty of the work. Other measures of quantitative work load included Hours Worked per Week, Hours of Overtime per Week, and Hours of Unwanted Overtime per Week — all self-report measures.

Variance in Workload refers to the extent to which the pace of the work remains steady or varies. In air traffic control, for example, there are 'rush hours' of traffic starting about 5 pm (17.00 hours).

Responsibility for Persons-E involves responsibilities for persons' futures, their job security, their wellbeing and their lives. By contrast, responsibility for things, not measured in this study, deals with responsibility for budget, equipment, and specific tasks or projects.

Job Complexity-E subsumes a number of diverse job characteristics: high contact with other people, task variation from one assignment to another, work on many different jobs in various states of completion at the same time, and the necessity of dealing differently with people from group to group.

Role Conflict refers to the presence of two or more conflicting demands from role senders or persons at work. The conflicting demands may come from one or several persons.

Role Ambiguity-E refers to uncertainty about what is expected of one on the job ('How often are you clear about what others expect of you?'). Job Future Ambiguity refers to certainty about job and career security ('How certain are you about whether your job skills will be of use and of value five years from now?').

Underutilization of Abilities refers to disuse of one's particular skills and abilities. Participation is defined as the amount of influence the person has on shared decisions which affect the person.

Inequity of Pay refers to the extent to which the person believes that there has been adequate compensation for the person's effort compared to reference groups such as others with similar skills within and outside the company. Virtually no one reported being overpaid.

The items on Social Support cover the extent to which people around the employee are good listeners or are persons on whom the employee can rely when help is needed ('How much does each of these people go out of their way to do things to make your work life easier for you?'). Support from the following three sources was measured: 'Your immediate supervisor', 'Other people at work', and 'Your wife, friends and relatives'.

Measures of Personality

The Sales Type A Personality subset (Vickerse, 1975) is based on traits of involvement in work, persistence, achievement orientation, a sense of time urgency, and being hard-driving. For example, 'I hate giving up before I'm absolutely sure that I'm licked'. The concept of a Type A behavior pattern that characterizes coronary-prone patients was developed by Friedman and Rosenman (see Rosenman *et al.*, 1970) and parallels earlier anecdotal and retrospective studies (see Dunbar, 1948). The Type B person is characterized as being the opposite of the Type A. Evidence of construct validity for the Sales measure is presented in Caplan and Jones (1975), and in Keenan and McBain (1979), and evidence of convergent validity is presented in Spicer (1980).

Flexibility (Gough, 1957) was measured by a scale that characterizes the person as one who tolerates words like 'approximately' or 'perhaps' and does not require everything to be in its place. The rigid person is the opposite.

Assert Good Self and Deny Bad Self are two seven-item factors derived by Lillibridge (1970) from the Need for Social Approval Scale (Crowne and Marlowe, 1964). Theoretically, asserters present themselves in a good light to win social approval, and deniers defend their self-concepts by under-reporting aspects of the self which are not socially approved. The items in the two factors derive from the same scale, but their correlation was only .45 for the random sample.

Measures of Person–Environment Fit

A number of studies and theories (Caplan, 1971; French, 1973; French, Rodgers and Cobb, 1974; Kulka, 1976; Harrison, 1976; Murray, 1938; Pervin, 1968) suggest that the fit between personality and job environment may be an important predictor of strain. As noted in Chapter I, the goodness of fit may explain variance in strain in addition to that accounted for either by personality or environment. Strain would result from discrepancies between either environmental demands and an individual's abilities to meet them or between an individual's needs and environmental supplies to meet those needs.

To measure the degree of fit between the respondents' perceptions of certain aspects of their environments and of themselves, two commensurate questions were used. An environment (E_s) item was paired with a person (P_s) item, for example, 'How much workload do you have?' and 'How much workload would you like to have?' Responses to both the E_s and P_s questions were marked on identical scales so they could be compared quantitatively.

Person–Environment (P–E) Fit was measured on eight dimensions. The preceding section concerning measures of the subjective environment included a description of the environmental (E_s) component of four multiple item measures of P–E Fit: Job Complexity-E, Responsibility for Persons-E, Role Ambiguity-E, and Workload-E. In addition to rating their jobs on each of these work characteristics, the men were asked to indicate the amount of each characteristic they would prefer to have on the job. Responses to both the E_s and P_s items were marked on Likert scales which typically ranged from $1 = $ 'very little', to $5 = $ 'A great deal'.

Four single item measures of P–E fit were obtained for the dimensions of income, overtime, length of service, and education. The E and P measures for equity of income are 'How much money did you earn . . .?' and 'How much do you think you should have been paid?' The E and P measures for overtime are '. . . how many hours of what you consider 'overtime' did you put in?' and 'How many of these overtime hours did you actually want to work . . .?' The E and P measures for length of service are 'How long do you feel a person needs to work in your particular job to be fully trained?' and 'How long have you been in your present position . . .?'. The E and P measures on education are 'What level of formal education do you feel is needed by a person in a job such as yours?' and 'How much schooling have you had?'.

To create a measure of P–E fit on a dimension, the discrepancy between the E and P measure was calculated. Two methods of calculating P–E fit scores were considered: a simple subtraction of the P score from the E score (E–P), itself computable in two different ways, described later, and a more complicated calculation using the ratio of the difference between the E and P scores compared to the level of the person's score ([E–P]/P). The rationale for each method of calculating P–E fit scores is presented in Chapter III, along with the summary of an analysis contrasting the utility of each method. The theoretical considerations and empirical results lead to the use of the difference formula in calculating P–E fit scores for job complexity, responsibility for persons, role ambiguity, workload, and overtime. The ratio formula was used to calculate P–E fit scores for income, length of service, and education.

For all of the P–E fit measures, a score of zero represents perfect fit (that is E equals P). A negative P–E fit score occurs when the person score is higher than the environment score. A positive P–E fit score occurs when the environment score is higher than the person score.

It has been argued that the relationship between P–E fit indices and strain may be linear or follow one of three curvilinear shapes (Caplan, 1971; French, 1973; French, Rodgers and Cobb, 1974; Harrison, 1976). The rationale for the various shapes of relationships is presented in the first part of Chapter III. To simplify the testing of the various relationships, transformations of the P–E fit measures were produced which would have linear relationships with strains when the underlying relationship followed one of the three predicted curvilinear shapes. To distinguish the original P–E fit measure from transformations of it, the original fit measures on a dimension is labelled the 'Fit' measure (for example, Workload Fit).

A U-shaped relationship exists between P–E fit and strain when both positive and negative P–E fit scores of increasing magnitude are related to increased strain. This U-shaped relationship becomes linear when the absolute value of the P–E fit scores is related to strain. The measures that result from the absolute value transformation are labelled 'Poor Fit' measures (for example, Workload Poor Fit) because strain is associated with poor fit in either direction.

Transformed measures were also developed for the instances where only negative discrepancies or only positive discrepancies are related to strain. When only negative P–E fit scores are related to strain, a linear relationship between the fit scores and strain is produced when all positive scores are reassigned the value of perfect fit, that is zero. The measures resulting from this transformation are labelled 'Deficiency' measures (for example Workload Deficiency) since the environment is deficient in matching the person scores. In such a case the person can also be said to have an excess of ability. When only positive P–E fit scores are related to strain, a linear relationship between the fit scores and strain is produced when all negative scores are reassigned the value of perfect fit. The measures resulting from this transformation are labelled 'Excess' measures (for example, Workload Excess) because the environment exceeds the person scores. In such a case the person can also be said to have a deficit of ability.

As indicated in Table II.3, the original 'Fit' measure and the 'Poor Fit', 'Deficiency', and 'Excess' transformations were produced for P–E fit on job complexity, responsibility for persons, role ambiguity, workload, length of service and education. As described below in Chapter III, scores on the original 'Fit' measure did not have both positive and negative values for P–E fit on income and overtime. It was rare to find employees with an excess of income or a deficiency of overtime. The transformations would therefore not alter the shapes of relationships for these two dimensions. So no transformed measures were produced for these two dimensions of P–E fit.

As indicated in Table II.3, four of the P–E fit measures are multiple-item indices. The reliability coefficients for the original multiple-item measures of 'Fit' ranged from .54 to .83. The reliability coefficients for the transformed multiple-item measures ranged from .54 to .88. Additional methodological considerations concerning the P–E fit measures are included in Chapter III.

Measures of Psychological Strain

Table II.3 presents the measures of psychological strain used in this study, their sources, and the cross-sectional reliabilities of the measures. The measures have all been used in previous studies where their reliabilities and validities were demonstrated. The range of reliability coefficients among these measures was from .75 to .86.

Three types of job satisfaction were measured. The first is labelled as Job Dissatisfaction because it refers to dissatisfaction without alluding to specific aspects of work. The measure has nonspecific items such as 'All in all, how satisfied would you say you are with your job?' Two measures tapped more specific reactions to the job: Workload Dissatisfaction and Boredom. Workload Dissatisfaction deals with how satisfied people are with their workload. The measure includes items such as 'I am unhappy about my current workload'. The Boredom measure contains items such as 'The work on my job feels dull' and 'I feel bored with the work I have to do'. The full item content and the inter-item correlations for these job satisfaction indices and for the other psychological strains appear in Appendix E of Caplan *et al.* (1980).

Indices of Somatic Complaints, Anxiety, Depression, and Irritation were also included in our questionnaire. As should be evident in Table II.3, several of the measures had multiple origins with each originator giving very similar items a slightly different twist in the format and phrasing. We wanted all the items measuring anxiety, depression, and irritation to have the same stem and response scales so that we would not end up with some indices (or even items within indices) being measured on four-point scales and some being measured on two, three or five-point scales. We also wanted to orient the items toward the measurement of states rather than traits. Trait anxiety would tap a personality disposition. We were interested in anxiety which varied as a function of the job environment. The stem was consequently worded to reflect how the person felt nowadays rather than how the person *generally* felt: 'When you think about yourself and your job nowadays, how much of the time do you feel this way?' (Spielberger, Gorsuch and Lushene, 1970). Similar wording of the stem was utilized in measuring somatic complaints so that we could tap symptoms which may have arisen only recently: 'Have you experienced any of the following during the past month . . . .?'

The measure of Somatic Complaints includes a variety of symptoms (such as sweating palms, upset stomach, loss of appetite, trouble sleeping, heart beating faster than usual) which have been observed in persons suffering from neuroses or from severe psychological stress (bankruptcy, having failed an exam, thinking about an upcoming evaluation of one's work, disasters, and so forth).

Thirteen items comprising three indices of affect were selected from nineteen items of the questionnaire on the basis of their intercorrelations and on rational grounds. The measure of Anxiety includes three items referring to the negative side of this feeling state (nervous, jittery, fidgety) and one referring to its absence (calm). The measure of Depression was derived from a factor analysis by Cobb (1970) of the Zung (1965) scale. The Depression measure contains both negatively worded items (depressed, sad, blue, unhappy) and positively worded items (good, cheerful). The three positively worded items were included in our measures even though the effect was to increase the correlation between the Anxiety and Depression indices ($r = .51$, $p < .001$). They were included because their scoring had to be reversed, and thus they served to reduce the proportion of response style variance in their indices (see Bentler, Jackson and Messick, 1971; Pinneau, 1973). The three negatively phrased items which tapped the dimension of Irritation (angry, aggravated, and irritated or annoyed) were combined to form the third affect.

Measures of Health-Related Behavior

Persons reported whether they were smokers, ex-smokers, or had never smoked as a habit. Smokers were asked to report the number of cigarettes, cigars, and pipes of tobacco smoked per day. Only cigarette smoking among smokers was coded because of the small number of pipe and cigar smokers.

Caffeine consumption per day was also measured by self-report. It included Cups of Coffee and all Caffeinated Drinks (the latter including coffee, tea and cola).

Obesity index ((weight in lbs/height in inches^2) $\times$ 100) was calculated according to formulas suggested by epidemiological studies (Florey, 1970; Goldbourt and Medalie, 1974). Self-reported height and weight were used with some confidence because of a study showing a correlation of .94 between actual and reported weight (Cobb, Tomlin, Chun, French and Kasl, 1974).

Measures of Illness

The reasons for the most recent dispensary visit were also asked. Obviously these self-reported diagnoses do not have the validity of medical records. Nevertheless, for some types of analysis they may approximate the relative distributions of illness in the samples.

Physiological Measures

Table II.4 lists the physiological measures used in the study and presents reliability data on each. The repeat reliability of the blood sera determinations was measurable because two blood specimens were drawn 20 minutes apart on 115 respondents. The measurement procedures are detailed in Caplan et al. (1980). All determinations were done at the Institute for Social Research[1] except where otherwise noted.

Blood pressure and serum cholesterol were measured because they are well known as risk factors in coronary heart disease. A paper by Stamler et al. (1969) suggested that heart rate may also be a risk factor and heart rate is a physiological determinant of blood pressure, so heart rate was also measured.

We decided to measure thyroid function because Levi (1972) noted its variation in relation to social stress. Mason (1968) notes that although thyroid function has been related to stress in a number of studies, the direction of the relationship varies. In some studies T3 and T4I increase; in other studies they decrease. The mechanisms behind this pattern of results is not understood. T3 and T4I should be positively correlated but, for unknown reasons, they were negatively correlated in our analyses ($r = -.28$).[2] So the results using these determinations have to be interpreted with caution.

Serum uric acid was examined because a number of studies show its association with stress and its tendency to vary by occupation in men (see, for example, Cobb, 1974; Dunn et al., 1963; Kasl, Cobb and Brooks, 1968;

---

1. They were performed by Mr Terry Roth under the supervision of Sidney Cobb, MD.
2. The T3 and T4I determinations were performed by a commercial laboratory because suitable equipment was not available to us.

Table II.4 Physiological measures

Measures	Abbreviations	Reliability: different samples 20 minutes apart		Reliability: same sample different days	
		n	r	n	r
Systolic Blood Pressure	Systolic BP	131	0.84	12[a]	0.99
Diastolic Blood Pressure	Diastolic BP	131	0.78	12[a]	0.99
Heart Rate	Heart Rate	127	0.81		
Serum Cholesterol	Cholesterol	111	0.91	65	0.98
Thyroid Hormone T3	T3	99	0.68	101	0.95
Thyroid Hormone T4I	T4I	99	0.77	101	0.97
Serum Uric Acid	Uric Acid	75	0.97	23	0.99
Serum Cortisol	Cortisol	96	0.70	19	0.83[b]

[a]London School of Hygiene Blood Pressure Test Tape.
[b]Correlation between values on different days. Each day's value is the mean of two determinations (correlations between determinations on the same sample on the same day are $r = .99$, $n = .43$).

Mueller and French, 1974). Even if occupational differences in uric acid exist, it is not known if these differences are a result of job stress or the result of the types of persons who are selected for the occupations.

The hormone levels of the adrenal cortex are particularly responsive to environmental stress (Selye, 1956). Accordingly, we decided to study adrenal cortical function (cortisol) in this study.[3]

Analysis Considerations when Using the Physiological Variables

The analyses, whenever possible, used the mean of multiple determinations of each person's pulse rate, blood pressure, or serum value. In this way, the reliability of each person's score was maximized.

Analysis of physiological variables in relationship to job demands were complicated by the presence of confounding variables. Circadian and seasonal rhythms were one such class of confounding variables, and age represented another class. Cortisol varied diurnally in our sample, whereas the thyroid hormones showed significant seasonal variation. The seasonal variation of T3 and T4I was particularly problematic because different occupations had their blood samples drawn at different times during the five-month period (eta^2 between occupation and Julian date was .70). Consequently, when one compares occupational groups on mean levels of the hormones, one also compares hormones collected at different times in the seasonal cycle and vice versa. While it is desirable either to collect all physiological data from all occupational groups during the same relatively short time-period, or else to distribute the collection over time equally for all persons in each occupational group, neither of these strategies was feasible in this study. The scheduling needs of the participating sites and the unexpected time needed to negotiate with reluctant or busy organizations added to the difficulty. As a result, the data on cortisol, T3 and T4I will need to be interpreted with caution.

Age was another potential confounding variable in the study of physiological variables. In some of our analyses, relationships between predictors and physiological dependent variables were examined within age-stratified groups as a control on the effects of age. In other analyses multiple regression was used to produce physiological measures adjusted for age and for other control variables. Table II.5 presents the results of these analyses which were conducted in part by Pinneau (1975). Pinneau conducted these analyses after determining that the control variables could account for variance in six of the eight physiological measures examined in a linear fashion, and after determining that there were no significant interactions between control variables in predicting to the physiological strains. Whenever the adjusted measures of the physiological strains are presented in tables in this report, they are labelled with an 'A' (such as Heart Rate-A).

---

3. The cortisol determinations were performed in the laboratory of Dr Robert M. Rose at Boston University Medical School under the supervision of Mr David Sandwisch.

Table II.5   Multiple regression of control variables on physiological strains

Strain	Control	Correlation or partial correlation[a]	Significance <	r^2 or R^2	Eta^2 (from Pin- neau, 1975)
Heart Rate ($N = 344$)	Smoking[b]	.41	.0001	.17	.17
Systolic BP ($N = 339$)	Obesity Smoking[b]	.28 .19	.0001 .0005	.11	.10
Diastolic BP ($N = 382$)	Obesity Age	.30 .10	.0001 .05	.11	.09
Serum Cholesterol ($N = 351$)	Age Obesity	.25 .19	.0001 .0005	.10	.14
Serum Uric Acid ($N = 337$)	Obesity	.29	.0001	.08	.06
Serum Cortisol ($N = 349$)	Time of day	−.44	.0001	.19	.12

Note: T3 and T4I have no significant relationships with potential confounding variables measured in the study. Data are from the physiological sample.

[a]When one control variable is regressed on a strain, their correlation is presented. When two control variables are regressed on a strain, the partial correlation between each control variable and strain is presented.

[b]Number of cigarettes smoked in the four hours preceding the data collection.

Methods of Analysis

Multivariate analyses were used including multiple regression and partial correlation, and tests for interaction effects using analysis of variance. A two-tailed test of significance was used unless otherwise noted. The analyses were conducted using the OSIRIS (Institute for Social Research, 1973) and MIDAS (Fox and Guire, 1976) software for data management and statistical analyses.

In many cases, special analysis procedures were used to test a particular hypothesis. These procedures are described in the text just prior to the presentation of the results of the particular analysis.

Chapter III

Person–Environment Fit

The introductory chapter presented a theoretical model of how characteristics of the job and the person affect strains and illness (see Figure I.1). In this chapter, we examine evidence that goodness of fit between the subjective person (P_s) and the subjective work environment (E_s) affect employee strain. For a more complete discussion of the theoretical and methodological issues associated with P–E fit and of the results presented in this chapter, see Harrison (1976).

The first section of the chapter reviews the theoretical principles which underlie the relationship between the person and the job environment. From this discussion, three hypothesized relationships between P–E fit and strain are developed. The second section of the chapter discusses the methodology of measuring P–E fit and analysing its effects on strain.

The third and largest section of the chapter presents the results of three sets of analyses exploring the relationship between P–E fit dimensions and strain. The first set of analyses examines the shape and strength of relationships between strain and P–E fit. The second set of analyses determines the amount of variance in strain accounted for by P–E fit measures beyond that accounted for by the component measures of the person and the environment. These results examine the extent to which predictions are improved by taking into account the predicted interactions between the person and the environment. The final set of analyses examines the variance in strain which is independently associated with various measures of stress. These results indicate the extent to which the measures of fit have direct causal relationships with strains.

Theory of P–E Fit and Strain

The utility of P–E fit theory is based on the assumption that people vary in their needs and abilities just as jobs vary in their incentives and demands. When there is a poor fit between the characteristics of the person and related characteristics of the job, P–E fit theory predicts that employee wellbeing will be reduced. This is the key proposition underlying the theoretical model presented in Chapter I (Figure I.1). The prediction presented above is too general to be of practical use in applied settings or of much interest to theoreticians. This section presents more specific hypotheses that deal with questions such as, 'Do all dimensions of the person and of the environment have the same effect on wellbeing?' and 'Are all indicators of wellbeing affected the same way by person–environment misfit?'

Different Forms of Person–Environment Fit

The first specification to be made is between two forms of person–environment fit. The differentiation of these two types of fit was introduced in Chapter I. One form of fit involves the discrepancy between the motives of the person and the supplies in the job and environment to meet the goals and preferences induced by those motives. Motives can be identified by reviewing the types of needs and values that have been studied. These include the motives to achieve, to gain power, to nurture, to succor, to tell the truth and so on. Lists of such motives can be found in the work of Murray (1938) and others (see, for example, Lawler, 1973; Pervin, 1968; Rokeach, 1973). A good fit between the person and the environment occurs when the supplies in the environment (for example, money, supportive people, opportunities to achieve) are sufficient to satisfy the motives of the individual.

A second form of fit deals with the relationship between the demands of the job and the abilities of the person to meet those demands. Consider the example cited in Chapter I concerning a secretary who has the ability to type only 60 words per minute and works in a job requiring a speed of 70 words per minute to handle the workload. This misfit might have a direct effect on the 'wellbeing' of the organization because the workload will usually not be handled. However, the misfit will not affect the wellbeing of the secretary unless it affects supplies to satisfy the secretary's motives. If the job and pay level are secure, the esteem from others is not affected by the discrepancy, and the individual's self-image is not tied to the discrepancy in work performance, then the discrepancy has no effect on supplies for motives and therefore should not affect the wellbeing of the secretary. If, however, supplies for any of these motives are threatened by discrepancies between the demands and abilities, then the individual's wellbeing should be affected.

Having considered the relationship of fit between demands and abilities to fit between motives and supplies, we now turn to the relationship of both kinds of fit to strain.

Shapes of Relationships between P–E Fit and Strains

Person–environment fit theory emphasizes the causal relationship between misfit and strain. The exact content and processes of the relationship remain unclear. Several factors are hypothesized to determine which strains will occur in response to poor P–E fit. They include (1) motives which are not being met, (2) abilities to meet demands of the job, (3) the genetic and social background of the individual, (4) defense and coping predispositions of the individual, and (5) situational constraints on particular responses. Predictions of which strain will occur in response to person–environment misfit require information concerning each of these five factors. In general, one can say that some forms of strain are more relevant to the misfit than others and that the most relevant dimension of strain will be the one which is most commensurate with the misfit. For example, a machine-paced operator who is required to produce ten pieces per hour but has an ability to produce a maximum of nine pieces per hour will certainly show the strain of dissatisfaction with workload.

Person–environment fit theory predicts that relevant strains should increase as P–E fit dimensions reflect increased insufficiency of supplies for motives. French, Rodgers and Cobb (1974) and Caplan *et al.* (1980) use the likely occurrence of insufficient supplies for motives to identify three potential shapes of relationships between P–E fit dimensions and strains. These relationships are illustrated in Figure III.1. The horizontal axis of the figure represents a scale of person–environment fit. The numbers on the scale

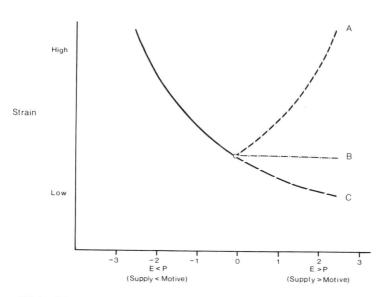

Figure III.1   Three hypothetical shapes of the relationship between P–E fit on motive–supply dimensions and strains

represent discrepancies between person and environment scores on commensurate dimensions. For example, the scale could represent the discrepancy between the individual's desire to interact with other people and the extent to which the job provides opportunities for the individual to work with others. The zero at the center of the scale represents the point of perfect fit where the person score and environment score are equal. The negative scores to the left of perfect fit represent a deficiency where the person score is increasingly larger than the environment score. In this instance, the individual would want to interact with others more than the job allows. The positive scores represent an excess where the environment scores are increasingly larger than the person scores. These scores would indicate that the individual must interact with others more than is desired. The vertical axis in Figure III.1 represents any strain which may result from sustained misfit such as job dissatisfaction or irritation with others.

The solid line in Figure III.1 illustrates the monotonic decrease in strain associated with the increase in environmental supplies to the point of matching the motive levels. This monotonic relationship applies to all motive–supply dimensions where supplies are not sufficient to meet motive levels. Insufficient food, money, love, opportunities for growth or any other supply will result in strain.

The relationship between fit and strain becomes complex when the effects of supplies in excess of motive strength are considered. Excess supplies may result in no change in the level of the strain, decreased strain, or increased strain depending on whether or not excess supplies for the motive alter supplies for other motives. These possibilities are considered below.

House (1972) points out that fit would have an asymptotic relationship with strain (described by the solid line and line B in Figure III.1) when excess supplies for one motive are not exchangeable for supplies for other motives. For example, a hungry person eats until full and then does not eat additional food even if it is offered. Similarly, the dimension of fit on opportunities for personal growth and development should be asymptotically related to strain. As a person is presented with more opportunities than the person wants or can use, the arousal of the motive for personal growth should remain about as low as it was when opportunities just met the need.

The monotonic relationship represented by the solid line and curve C in Figure III.1 will occur when excess supplies for one motive can be used directly as supplies for other motives. One of the most common uses for excess amounts of *preservable* supplies is to store them in case the availability of the supplies should be reduced in the future. The accumulation of preservable supplies for various motives is in itself a goal of security motives. The monotonic relationships represented by the solid line and curve C in Figure III.1 will also occur when excess supplies for one motive can be *exchanged* for supplies for other motives. For example, additional salary from overtime work can be used to purchase supplies for other needs. By satisfying other motives, strain is reduced below its level at perfect fit on income.

The reader should note that a monotonic relationship which is not the same as curve C may occur between certain measures of fit and strain. In situations where the supplies do not exceed the level of motive strength, fit scores will fall only along the solid line in Figure III.1. For example, Evans (1969) and Wanous and Lawler (1972) found that most people reported that their jobs do not offer sufficient supplies for their higher-order needs. Therefore, only deficiency scores were found on this dimension of P–E fit. These authors reported linear relationships between these deficiency scores and job dissatisfaction that were similar to the solid line in Figure III.1.

The final relationship between P–E fit and strain is the U-shape illustrated by the solid line and curve A in Figure III.1. French *et al.* (1974) suggest that the presence of excess supplies for one motive may result in deficient supplies for another motive. Consider, for example, P–E fit on the dimension of privacy on the job. Strain should be high when the individual has less privacy than is wanted, with strain decreasing as the opportunities for privacy increase toward the desired level. As the individual experiences more privacy than is wanted, supplies for affiliation motives may be reduced thereby increasing strain.

The quality of the specific strain responses may differ from one side of the U-shaped curve to the other because the two sides represent insufficient supplies for different motives. For example, too little privacy may lead to feelings of irritation with others who intrude on the individual. Too much privacy may result in feelings of loneliness and boredom. Both types of responses should be associated with overall job dissatisfaction. Job dissatisfaction should therefore increase with both deficiencies and excesses on the dimension of P–E fit on privacy, whereas irritation should increase only with deficiencies on the dimension, and loneliness should increase only with excesses on the dimension.

The relationships between fit on demand–ability dimensions and strain are similar to those between fit on motive–supply dimensions and strain because both sets of relations are based on the extent to which motives may be satisfied. The only difference between these two sets of relationships is the reversed meaning of the P–E fit dimension when P–E fit continues to be defined as the person score subtracted from the environment score (P–E fit = E – P). When job demands (such as typing speed, work pace, leadership, etc.) effect on supplies for motives and if so, would not affect strain. For example, having a knowledge of mathematics somewhat beyond that required threatened and strain should occur. The strain should increase monotonically as job demands increase beyond worker ability. Therefore, strain should always increase with positive P–E fit scores on dimensions reflecting demands and abilities.

Negative P–E fit scores on dimensions reflecting demands and abilities indicate that the individual's abilities exceed the demands of the job. The possible effects of excess abilities on strain parallel the three possible effects of excess supplies on strain. When excess abilities are present they may have no effect on supplies for motives and if so, would not affect strain. For example, having a knowledge of mathematics somewhat beyond that required by the job may have no effect on supplies for other motives, such as the need for prestige, for variety in work, or for social interaction. Alternatively, excess abilities for some job demands may provide supplies for other motives, reducing overall strain below the level of strain for perfect fit on job demands. For example, being able to handle easily the required workload may allow time for socializing, reading or other activities which satisfy other motives and increase overall job satisfaction. Finally, excess abilities may indicate that important motives, such as the need for self-utilization, are not being met and so strain should increase. For example, an individual with abilities to handle a job much more complex than the present one may value these abilities as a part of his or her self-identity. The inability to use those abilities could result in lowered self-esteem and other strains.

Hypothesized Relationships in the Present Study

What relationships are expected between strains and P–E fit measures in the present study? Harrison (1976) developed a detailed set of predictions for relationships between each P–E fit measure and strain. The predictions for relationships were derived by considering the motives likely to underlie misfit on each P–E fit dimensions and the circumstances likely to evoke the various types of strain. The detailed set of predictions are too lengthy to review here, but some summary observations are appropriate.

Six of the eight dimensions of P–E fit measured in the present study are dimensions indicating fit between specific characteristics of the job, and the preferences of the individual. These six dimensions are job complexity, role ambiguity, responsibility for persons, workload, overtime, and income. When the preferences or goals of the individual are below the level of these demands, one or more types of strain should occur. Additionally, strains may occur when the individual's preferences exceed the demands of the job. Strains may occur less frequently and be of lesser magnitude when the demands are insufficient than when the demands are excessive.

Two additional dimensions of P–E fit included in this study are education and length of service. These dimensions reflect the discrepancies between the education and time on the job usually required to perform the job well and the actual education and length of service of the worker. Discrepancies on education and length of experience are assumed to reflect typical or likely discrepancies between demands and abilities. Less education or length of service than is typically required may reflect insufficient general abilities and general job knowledge. More education or experience than is typically required may reflect more than adequate abilities and knowledge. Education and length of service may reflect abilities but do not measure discrepancies concerning specific job demands. Consequently, these two fit dimensions may show a different pattern of correlations between strains and P–E fit than shown by the fit dimensions more clearly specifying job demands and individual motives and preferences.

Methodological Considerations

Before the relationship between P–E fit dimensions and strains could be analyzed, three methodological activities had to be performed. First, the distributions of the P–E fit measures had to be reviewed to be sure the data were adequate to test the hypothesized relationships with strains. Second, transformations of the P–E fit dimensions had to be produced that would have linear relationships with strains if the original relationships were asymptotic or U-shaped. Finally, alternative procedures for calculating P–E fit scores had to be examined to select the procedure best supported by theoretical considerations and empirical findings. These three steps and their results are detailed below.

Distributions of the P–E Fit Measures

Three potential problems with the distribution of P scores and E scores in a sample can restrict one's ability to perform analyses to test for the predicted shapes of relationships between P–E fit and strain. One problem arises if the relative sizes of the E and P scores differ consistently. If the E scores are consistently greater than the P scores, the range of P–E fit scores will be limited to only positive values. Similarly, E scores being constantly lower than P scores limit the range of P–E fit scores to only negative values. In either instance, tests of the curvilinear relationships between P–E fit and strains are not possible because the scores will not be distributed on both sides of the turning point represented by perfect fit.

A second methodological problem with the distribution of E and P scores occurs when the variance of either the E or P scores is small compared to the variance of the other. The variance in the P–E fit scores then reflect mostly the variance in the component with the larger variance.

A third methodological problem which can affect P–E fit scores is a high correlation between the E and P scores. As the scores on component measures become more highly correlated, more individuals have the same P–E fit scores. When the correlation is very high, the variance in the P–E fit scores may be so limited that relationships with strains cannot be observed within the sample. For P–E fit on job-related dimensions, a moderate level of correlation occurs because of the tendencies for individuals to seek out jobs which meet their interests, for job environments to socialize the individual to better fit with the job, and for psychological defenses to improve the subjective perception of the fit between the person and the environment. These processes increase the likelihood that the E and P component scores for an individual will be at the same level, thereby increasing their correlation and the number of P–E fit scores at and around perfect fit.

Preliminary analyses were performed to determine the extent to which these three methodological problems would limit the effectiveness of tests for the predicted shapes of relationships in our data set. Table III.1 presents data concerning the restriction of P–E fit scores in the random stratified sample to all positive or all negative scores. Most dimensions of P–E fit exhibited good distributions on both sides of perfect fit. However, restricted distributions are evident in Table III.1 for the measures of P–E fit on income and overtime.

Table III.1 Distribution of scores on P–E fit variables

P–fit on	Standard deviations from perfect fit[a]/%							SD on original scale
	−3 or less	−2	−1	0 Perfect fit	1	2	3 or more	
Job Complexity	4	4	13	57 (22)[b]	17 4	4	1	.87
Role Ambiguity	3	7	19	46 (15)	21	4		1.24
Responsibility for Persons	3	15	28	40 (12)	11	3		1.27
Workload		1	15	48 (9)	23	10	3	.64
Income	10	23	34	33 (31)				.11
Overtime				84 (80)	7	4	5	10.3
Length of Service		10	30	42 (29)	12	5	1	.43
Education	5	4	15	67 (56)	8	1		.21

Note The distributions of scores are expressed as the percentage falling within one or more standard deviations from perfect fit. The distribution may be expressed in the original scale of a variable by multiplying the standard deviations from perfect fit (that is the column headings) by the standard deviation of the variable scores on the original scale (that is entries in the last column).

The distribution is based on approximately 300 men in the random stratified sample.

[a]Each standard deviation interval includes scores within $\pm \frac{1}{2}$ SD of the column heading.

[b]Parentheses enclose the percentage of scores which exactly equal perfect fit (that is $E - P = 0$).

For income, individuals reporting a discrepancy between E and P scores invariably indicated they wanted more income than they had received. For overtime, individuals reporting a discrepancy invariably indicated they wanted less overtime than they had worked. These restrictions in range did not allow us to test for asymptotic and U-shaped relationships between strains and P–E fit on either income or overtime.

Table III.2 presents data concerning the variance of E and P scores and their relative contribution to the variance of P–E fit scores. The means and standard deviations of the E and P components for the eight dimensions in the random stratified sample are of sufficient size and are similar enough in magnitude to suggest that one component will not produce most of the variance in a P–E fit measure. The standard deviations for the respective E and P components in the physiological sample are similar to those presented in Table III.2.

The final problem to be checked was the extent to which E and P scores were correlated. The correlations between commensurate E and P measures in the random stratified sample were .13 for workload, .18 for role ambiguity, .26 for length of service, .42 for responsibility for persons, .69 for job complexity, .75 for education, .89 for overtime, and .93 for income. An inspection of a

Table III.2 Means and standard deviations of the E and P component measures

Dimension	E Mean	SD	P Mean	SD
Job Complexity	4.75	1.13	4.80	1.05
Role Ambiguity	2.01	.83	2.13	1.08
Responsibility for Persons	2.84	1.27	3.52	1.07
Workload	3.56	.53	3.31	.45
Income	17,379	11,382	19,692	11,108
Overtime	18.5	25.1	15.0	23.7
Length of Service	3.6	1.4	4.3	1.3
Education	13.5	4.0	14.4	2.9

Note   The means and standard deviations presented are for the random stratified sample.

two-way distribution for the P and E components on income showed that individuals typically wanted income levels slightly higher than they were receiving. Similarly, for overtime, individuals wanted either no overtime or levels slightly lower than they had. The high number of identical P–E fit scores on these dimensions may attenuate the relationships observed between these dimensions and strains in this sample.

In summary, the distributions of the commensurate E and P measures in the study appeared adequate and were generally not expected to restrict the ability of derived measures of P–E fit to relate to strains. The income and overtime dimensions had the most limitations. Fit scores on these dimensions were distributed on only one side of perfect fit.

Calculating P–E Fit Scores

Discrepancy theory and equity theory present alternative approaches for combining E and P scores to produce P–E fit scores (Lawler, 1973). Discrepancy theory suggests that the degree of fit is reflected by the algebraic difference of the scores (in other words, E–P). This computational formula focuses on the number of units separating the E and P scores. Equity theory suggests that the degree of fit is reflected by the ratio comparing the difference between the E and P scores to the size of the P score (that is, [E–P]/P). This computational formula focuses on the relative magnitude of the discrepancy in relation to the individual's preferences.

The different emphases of the discrepancy and equity formulas can be easily demonstrated with an example. Suppose that one individual receives $10,000 per year and feels that the salary should be $15,000, whereas another individual receives $45,000 and feels that his salary should be $50,000. The discrepancy formula suggests the two individuals are experiencing an equivalent deficiency of $5000. The equity formula suggests the first individual is experiencing considerably more misfit than the second individual, a misfit of 33 per cent compared to a misfit of 10 per cent, respectively.

It can be argued that the best formula for combining E and P scores depends on whether an interval scale or a ratio scale is used to measure the E and P values. Many times, interval scales are used to measure the E and P values on several dimensions. For example, Likert scales with values ranging from 1 = 'very little' to 5 = 'a great deal' are assumed to approximate interval scales and have been used in the present study to measure varibles such as workload. Ratio scales (that is, scales with a zero point) are used to measure E and P values on other dimensions. For example, income is measured in dollars, overtime in hours, and length of service in months and years.

Values on ratio scales have meaning in proportion to one another. For example, five hours is five times more than one hour. The proportional relationship assumed by the equity formula is likely to be useful in calculating misfit when the E and P scores are measured on a ratio scale. However, values on interval scales do not have such proportional relationships. A score of '5' on a Likert scale cannot be said to be five times greater than a score of '1'. The interval relationship assumed by the discrepancy formula is likely to be most useful in calculating misfit when the E and P scores are measured on an equal interval scale (or in this case, a scale assumed to have equal intervals).

Pilot analyses were performed to determine which method of calculating P–E fit scores produces fit measures most strongly associated with strains. These analyses are detailed in Harrison (1976), and the results are summarized below.

Four P–E fit dimensions were measured using interval scales: job complexity, role ambiguity, responsibility for persons, and workload. For these dimensions the scores produced by the discrepancy formula almost always had slightly stronger correlations with strains than scores produced by the equity formula.

Four P–E fit dimensions were measured using ratio scales: income (dollars), overtime (hours), and length of service and education (both in years). For P–E fit on income, length of service, and education, the scores produced by the ratio formula almost always had slightly stronger correlations with strains than scores produced by the discrepancy formula. However, for P–E fit on unwanted overtime, the discrepancy formula produced scores which usually correlated more strongly with strains than scores produced by the ratio formula.

A conceptual problem with the dimension of unwanted overtime may have produced results for this dimension which differ from those of the other dimensions measured using ratio scales. Individuals may judge their fit on work hours in terms of total hours rather than on unwanted overtime only. If this is the case, the unwanted overtime measure uses the wrong P value in calculating the degree of misfit. The predictive power of the ratio measure of fit would be reduced since an incorrect value of P is divided into the discrepancy between E and P. This possibility and the better empirical performance of the discrepancy measure led to the decision to use the discrepancy measure for P–E fit on overtime in the analyses for the present study.

For the remaining P–E fit dimensions, both the theoretical differences in response scales and the empirical performance of the discrepancy and ratio measures led to the following decisions. Discrepancy measures were used for P–E fit on Job Complexity, Role Ambiguity, Responsibility for Persons, and Workload. Ratio measures were used for P–E fit on income, length of service, and education.

Before concluding the discussion of the two computational formulas utilized, it must be noted that the distinction between them is primarily theoretical. The choice between the discrepancy and ratio formulas for calculating P–E fit scores will have only a small effect on the empirical findings. Indeed, the P–E fit scores produced by the two formulas usually correlated .90 or higher. The point is made more generally by Nunnally (1967). Until the methods of measuring the E and P scores become more precise, the choice between the computational formulas will not markedly affect one's findings.

Transformations of P–E Fit Measures

The hypothesized relationships between P–E fit and strains include relationships that are approximately linear, asymptotic, and U-shaped. Many of the most powerful and commonly used analysis techniques (such as correlation and regression) examine linear relationships between variables. These analysis techniques can be used for nonlinear relationships if the P–E fit dimensions are transformed so that asymptotic or U-shaped relationships to strain become approximately linear. No transformation of the P–E fit scores is necessary when P–E fit is assumed to be linearly related to strain.

Transformation of U-shaped relationships. When the theoretical relationship between P–E fit and strain is U-shaped, strain increases as P–E fit becomes either greater than zero or less than zero. This relationship can be made approximately linear by taking the absolute value of the P–E fit scores. This transformation produces a scale of the magnitude of poor fit with no distinction made as to whether the discrepancy results from deficient or excess levels. Strain increases as the magnitude of poor fit increases. To distinguish the original P–E fit measure from transformations of it, the original P–E fit measure on a dimension will be referred to as the 'Fit' measure (for example, Workload Fit). Indices produced by taking the absolute value of P–E fit scores will be referred to as 'Poor Fit' indices (for example, Workload Poor Fit).

When this transformation is performed on a multiple-item index of P–E fit, it is possible to transform either the item discrepancies and then sum them or to sum the item discrepancies and then transform them. The two procedures will produce different results when the index contains items with both positive and negative scores. The theoretical and empirical usefulness of each transformation sequence were considered in Appendix D of Caplan *et al.* (1980). The more useful transformation sequence was generally to take the absolute value of the discrepancy for each item pair and then sum those absolute values. This procedure was used to produce the 'Poor Fit' transformation used in this study.

The absolute value transformation is appropriate for all P–E fit measures with scores on both sides of perfect fit. The income and overtime fit measures, however, have P–E fit scores on only one side of perfect fit. No 'Poor Fit' measures were constructed for these two dimensions.

Transformation of asymptotic relationships. The other situation requiring a transformation of P–E fit scores is when P–E fit is asymptotically related to strain. Strain increases as P–E fit scores increase in magnitude on one side of perfect fit (that is, either greater than or less than perfect fit) and strain remains the same as P–E fit scores increase in magnitude on the other side of strain. House (1972) suggests a simple transformation of the P–E fit measure to make the relationship approximately linear: the P–E fit scores on the side of perfect fit thought not to be related to strain are recoded to the value of perfect fit. This recoding assigns the same value to all cases where the level of P–E fit should produce no strain. Increases in strain should then be monotonically related to increases in the magnitude of the remaining scores on P–E fit.

Two asymptotic relationships between P–E fit and strain are possible, and therefore two transformed measures are necessary. One measure will have linear relationships to strain when only deficiency scores (that is, where $P > E$) are related to strain. In this case, excess scores (that is, where $P < E$) are reassigned values of perfect fit. This transformation of a P–E fit measure will be referred to as a 'Deficiency' measure (for example, Workload Deficiency). The other measure will have linear relationships to strain when only excess scores are related to strain. In this case, deficiency scores are reassigned values of perfect fit. This transformation of a P–E fit measure will be referred to as an 'Excess' measure (for example, Workload Excess).

When these transformations for asymptotic relationships are performed on a multiple-item index, it is possible either to transform the item discrepancies and sum them or to sum the item discrepancies and then transform the sum. These alternative procedures parallel exactly the alternatives for transforming multiple-item P–E fit measures with U-shaped relationships to strains. All of the theoretical conclusions made in Appendix D of Caplan *et al.* (1980) apply to this choice of procedures as well. Unfortunately, the transformations for asymptotic relationships in this report were performed on the index score rather than on the item scores. However, the transformation procedure should not greatly affect the relationships of measures of 'Deficiency' and 'Excess' with other variables. The items in the indices are rather highly intercorrelated (see the inter-item reliability estimates presented in Table II.3), indicating that either transformation sequence will produce similar results.

The 'Excess' and 'Deficiency' transformations are not appropriate for the income and overtime dimensions because they have P–E fit scores on only one side of perfect fit.

Obviously, the various fit measures for a dimension are interrelated because they represent different transformations of the same 'Fit' measure. The magnitude of correlation between measures on the same dimension is affected by the underlying distribution of P–E fit scores. For example, the distribution of Responsibility for Persons-Fit in Table III.1 shows that 45 per cent of the individuals reported deficient fit scores one or more standard deviations below perfect fit, and only 14 per cent of the individuals reported excess fit scores one or more standard deviations above perfect fit. The Deficiency measure, therefore, had much more overlap with the Poor Fit measure than did the Excess measure. Consequently, Responsibility for Persons-Poor Fit correlated –.83 and .10 with Deficiency and Excess measures on the same dimension.

In summary, the original P–E fit measure and the transformations discussed above were used to produce four measures for each dimension: Fit, Deficiency, Excess, and Poor Fit. By linearly relating each of these measures to strain, the shape of the relationship between a P–E fit dimension and a strain was checked for the linear, asymptotic, and U-shaped relationships predicted by the P–E fit theory.

Relationships between P–E Fit Measures and Strains

As was noted at the beginning of this chapter, three sets of analyses were performed to explore the relationships between P–E fit measures and strains. The first set of analyses produced correlations between the transformed measures of P–E fit and strain to examine the strength and shape of the relationships. The second set of analyses ascertained the amount of variance in strains accounted for by curvilinear transformations of P–E fit measures beyond that accounted for by the E and P component measures. The last set of analyses determined the variance in strain which was independently related to various predictors.

Correlations between Strains and Measures of P–E Fit

Tests of the hypothesized relationships between P–E fit dimensions and strains were performed by correlating the various P–E fit measures with the psychological, behavioral, and physiological strains. The transformed 'Deficiency', 'Excess', and 'Poor Fit' measures permitted the use of linear correlations to identify hypothesized asymptotic and U-shaped relationships between P–E fit dimensions and strains. The shapes of relationships represented by correlations using the different P–E fit transformations are summarized in Table III.3.

The correlations between fit measures and both psychological and behavioral strains were computed using the random stratified sample. The correlations between fit measures and physiological strains were computed using the physiological sample.

Table III.3   Summary of shapes of relationships represented by correlations between types of measures of P–E fit and strains

Measure	Sign of correlation	Shape of underlying relationship with P–E fit	
'Fit'	+	/	
'Fit'	–	\	
'Deficiency'	+	⌐/	a, b
'Deficiency'	–	_	a
'Excess'	+	_/	a
'Excess'	–	⌐\	a, b
'Poor Fit'	+	V	a
'Poor Fit'	–	∧	a, b

[a]The turning points on this shape are at perfect fit, that is, E –P $= 0$.

[b]This shape has not been hypothesized to occur. It suggests that strain would decrease as the fit between the person and the environment becomes worse.

The correlations between the eighteen strains and the E, P, Fit, Deficiency, Excess and Poor Fit measures for each of the eight P–E fit dimensions are detailed in Appendix C, pages 134–141. These results are highlighted in Table III.4. This table presents the strongest significant correlation between each strain and a measure of P–E fit on each dimension.

P–E fit on job complexity. The correlations between strains and P–E fit on job complexity indicate that strain was likely to occur when the job was either too complex or too simple compared with the individual's preference. The correlations with 'Poor Fit' on Job Complexity indicate that Job Dissatisfaction, Workload Dissatisfaction, Boredom, and Depression all increased when the job was too simple or too complex. A review of the correlations with each variant of the Job Complexity measure of P–E fit in Appendix C's data suggests that too much and too little complexity contributed equally to the experience of Job Dissatisfaction, that too much complexity contributed more strongly to Workload Dissatisfaction, to Depression and to Anxiety than did too little complexity, and that too little complexity contributed more strongly to Boredom than did too much complexity. Irritation was related only to too much complexity. Somatic Complaints and Number of Smokes were related only to too little complexity.

The correlations between P–E fit on job complexity and physiological measures only partially support the hypothesized relationships. The thyroid hormone, T3, did increase with too little complexity. However, the negative correlation between Job Complexity Poor Fit and the adjusted measure of heart rate was opposite to the direction hypothesized. The correlation suggests that heart rate decreases as the job becomes too complex or too simple. These relationships with the physiological variables barely reach the .05 criterion for inclusion (two-tailed test), so their reliabilities are open to question.

P-E fit on role ambiguity. The correlations between strains and P-E fit on role ambiguity suggest that jobs which are either too rigidly defined or too loosely defined are related positively to strain. Job Dissatisfaction, Workload Dissatisfaction, Boredom and Depression had U-shaped relationships with P-E fit on role ambiguity. A review of the correlations in Appendix C (pages 134–141) suggests that too much ambiguity made the primary contribution to the relationships with these strains. Excess role ambiguity was also positively related to Irritation.

The negative correlations with the adjusted measures of blood pressure (both systolic and diastolic blood pressure) were opposite to the direction expected. These correlations showed that blood pressure was lower as fit on Role Ambiguity was worse, particularly with Excess Role Ambiguity. These relationships cannot be easily dismissed. Although they are relatively weak, they equal the magnitude of some of the relationships with psychological variables and almost reach the .01 level of significance.

P-E fit on responsibility for persons. The correlations between strains and P-E fit on responsibility for persons indicate that both too much and too little responsibility of this type is related to strain. Overall Job Dissatisfaction and Boredom were related to both too much and too little responsibility for persons. The correlations in Appendix C suggest that too little responsibility affected Job Dissatisfaction and Boredom more strongly than did too much responsibility. However, Workload Dissatisfaction and Anxiety increased only with too much responsibility for persons.

P-E fit on workload. The correlations between strains and P-E fit on workload indicate that strain is likely to occur when there is either too much or too little work to be done. Job Dissatisfaction and Boredom correlated most strongly with Workload Poor Fit. A review of the correlations in Appendix C suggests that excess work had a stronger positive relationship with Job Dissatisfaction than did too little work. Excess work was also related positively to Workload Dissatisfaction, Anxiety, and Irritation, and it accounted for most of the relationship between Workload Fit and Depression.

The relationships between fit on workload and measures of physiological strain were weak, with only two significant correlations found. The positive correlation between Workload Fit and adjusted systolic blood pressure is in the expected direction and follows the pattern found for psychological strains. Blood pressure increased with excess workload and tended to decrease with too little work. Workload was also related to T3 — T3 decreased as excess work increased.

P-E fit on overtime. The correlations between strains and P-E fit on overtime follow the expected pattern with excess overtime relating to increased strain. There were positive correlations between Overtime Fit and Job Dissatisfaction, Workload Dissatisfaction, Anxiety, and Somatic Complaints. The

Table III.4   Strongest significant correlation

Dimensions

Strains	Job complexity[a]	Role ambiguity[a]	Responsibility for persons[a]
Job Dissatisfaction	.47 Poor Fit	.19 Poor Fit	.23 Poor Fit
Workload Dissatisfaction	.36 Poor Fit	.13 Poor Fit	.17 Excess
Boredom	.51 Poor Fit	.17 Poor Fit	.32 Poor Fit
Depression	.22 Poor Fit	.12 Poor Fit	NS
Anxiety	.21 Excess	NS	.12 Excess
Irritation	.19 Excess	.14 Excess	NS
Somatic Complaints	−.19 Deficiency	NS	NS
Number of Cigarettes Smoked	−.20 Deficiency	NS	NS
Coffee and Tea	NS	NS	NS
Obesity	NS	NS	NS
Heart Rate (A)[e]	−.11 Poor Fit	NS	NS
Systolic Blood Pressure (A)	NS	−.14 Poor Fit	NS
Diastolic Blood Pressure (A)	NS	−.15 Excess	NS
Cholesterol (A)	NS	NS	NS
T3	−.11 Deficiency	NS	NS
T4I	NS	NS	NS
Serum Uric Acid (A)	NS	NS	NS
Cortisol (A)	NS	NS	NS

Note The physiological sample (n = 390) was used to determine correlations with the eight physiological measures. The random stratified sample (n = 318) was used to determine the other correlations.

Correlations between strains and E, P, and P–E fit measures on each dimension are presented in Appendix C.

[a]Fit = E–P; Deficiency = E–P for E–P $\leq$ 0 and Deficiency = 0 for E–P > 0; Excess = E–P for E–P $\geq$ 0; and Excess = 0 for E–P < 0; and Poor Fit = |E–P|.

[b]Fit = E–P; no transformations were produced because E $\geq$ P.

[c]Fit = (E–P)/P; no transformations were produced because E $\leq$ P.

[d]Fit = (E–P)/P; Deficiency = (E–P)/P for (E–P) $\leq$ 0 and Deficiency = 0 for (E–P)/P > 0; Excess = (E–P)/P for (E–P) > 0 and Excess = 0 for (E–P) < 0; and Poor Fit = |E–P|/P.

[e]The expression '(A)' at the end of a physiological measure indicates the measure has been adjusted to remove the effects of extraneous variables such as age. See Table II.5 for the specific adjustments made to each physiological variable.

between each strain and each dimension of P–E Fit

		Dimensions		
Workload[a]	Overtime[b]	Income[c]	Length of service[d]	Education[d]
.22 Poor Fit	.19 Fit	NS	−.16 Deficiency	−.30 Deficiency
.54 Excess	.38 Fit	−.23 Fit	−.12 Fit	.19 Poor Fit
.12 Poor Fit	NS	NS	−.36 Deficiency	−.51 Deficiency
.27 Fit	NS	−.12 Fit	NS	.22 Poor Fit
.15 Excess	.18 Fit	NS	NS	NS
.29 Excess	NS	NS	NS	NS
NS	.13 Fit	−.17 Fit	NS	−.12 Deficiency
NS	NS	NS	NS	.20 Poor Fit
NS	NS	NS	NS	NS
NS	NS	NS	NS	NS
NS	NS	NS	NS	NS
.11 Fit	NS	NS	NS	NS
NS	NS	NS	NS	NS
−.11 Excess	NS	NS	NS	NS
NS	NS	NS	.11 Fit	NS
NS	NS	−.14 Fit	−.14 Fit	NS
NS	.12 Fit	.14 Fit	NS	NS

measure used to obtain overtime information made it impossible to record deficiency scores on overtime. Therefore, the positive correlations result entirely from the pattern of excess scores.

The one significant correlation with a physiological measure fits the hypothesized pattern. The adjusted level of cortisol increased with excess overtime.

P–E fit on income. The correlations between strains and P–E fit on income generally followed the expected pattern. Too little income was associated with increased strains. Income–Fit was negatively correlated with Workload Dissatisfaction, Depression and Somatic Complaints. No excess scores were reported for this P–E fit dimension, so the negative correlations result entirely from the pattern of deficiency scores.

Again, the two relationships with physiological strain were weak and inconsistent. The negative correlation between Income–Fit and adjusted levels of serum uric acid fits the hypothesized relationship and follows the pattern found for psychological strain. However, the positive correlation of Income Fit with adjusted cortisol level suggests that cortisol increases as fit on income improves.

P–E fit on length of service. The correlations between strain and P–E fit on length of service indicate that strains occur primarily when the person has more time on the job than is usually required to do the job well. Job Dissatisfaction and Boredom both increased with excess time on the job. Workload Dissatisfaction was negatively correlated with Length of Service-Fit. This relationship indicates that workload dissatisfaction is lowest when the individual has little experience on the job and increases for individuals who have more experience than is required to do their jobs.

Length of service had significant relationships with two physiological strains. The adjusted score for serum uric acid was negatively correlated with Length of Service Fit, following the same pattern as Workload Dissatisfaction. The positive correlation between T4I and Length of Service Fit had the opposite relationship, with levels of T4I decreasing for the persons with greater length of service than the level needed to perform the job well.

P–E fit on education. The correlations between strains and P–E fit on education indicate that strains occurred primarily when the person had more education than was typically required to perform the job. Job Dissatisfaction, Boredom, and Somatic Complaints increased with too much education. A review of the correlations in Appendix C suggests that the correlations of Education-Poor Fit with Workload Dissatisfaction, Depression, and Number of Cigarettes Smoked resulted primarily from the relationship of too much education with strain.

Discussion of Correlational Results

The overall results presented in Table III.4 provide many instances of the asymptotic and U-shaped relationships predicted by the P–E fit theory. The shape of a particular relationship depends on the dimension of P–E fit and the strain being examined. Some characterization, however, of the overall pattern of results is possible. The shapes of relations for the P–E fit dimensions is considered first, and then the pattern of relationships for types of strains is considered.

In formulating predictions concerning the shapes of relationships between measures of P–E fit and strain, five dimensions were identified as indicating fit between demands of the job and the preferences of the individual. These five dimensions head the first five columns of Table III.4: Job Complexity, Role Ambiguity, Responsibility for Persons, Workload, and Overtime. A fairly consistent pattern emerges from the significant relationships between strains and P–E fit on these dimensions.

Excessive job demands were associated with increased levels of most strains. This pattern is particularly striking when the psychological and behavioral strains are considered: 25 of 27 significant relationships showed strain increasing with excessive demands. Additionally, fifteen of the 23 significant relationships with psychological and behavioral strains showed strain increasing with insufficient demands. (The four significant correlations with P–E fit on overtime did not test this relationship because no deficiency scores were obtained for this dimension.)

P–E fit on income reflects fit between supplies from the job and preferences of the individual. As expected, insufficient income was associated with increasing strains. The absence of reports of too much income precluded tests of the predicted relationship between it and lowered strain.

Of the eleven significant correlations between strains and P–E fit on Length of Service or on Education, ten showed strains increasing as experience exceeded that required to do the job well (in other words, the job required less experience or education than the person had). The strength of the relationships between strains and P–E fit on length of service and on education was unexpected. These correlations were generally of the same magnitude as the correlations with the more specific dimensions of job demands. Those strains that were related to Length of Service and Education are primarily ones that were correlated with being overly experienced or overly educated for the job. In addition to misfit on specific job demands (such as, Job Complexity), misfit on Length of Service and Education may reflect stresses resulting from being locked into a 'dead-end' job or from being underemployed. Thus, a lack of fit between the nature of the job and the person and the inability to leave the job may explain the relative strength of these correlations.

Only four of the eleven significant relationships showed strains increasing as the individual had less experience than was typically required to do the job well (in other words, the job required more length of service or education than the person had). The weak effect of being underexperienced may result in part from a bias in the sample chosen for this study. The focus of this study is chronic job stresses, so only individuals with at least six months' experience in their jobs were included. This sampling approach therefore excludes individuals with less than six months on the job. Furthermore, it excludes individuals whose lack of education sufficiently hampered their performance to the extent that they were removed from their positions within the first six months on the job. Eliminating individuals not retained for at least six months may have weakened the relationship between strains and the extent to which the person has either less length of service or less education than is typically required to perform well.

Before concluding the discussion of the correlational analyses, the differing results for the different types of strain should be noted. The expected relationships were most often found and were strongest for the psychological strains and reported somatic complaints. Individuals' affective psychological reactions were evidently most closely associated with their perception of fit between themselves and their job. The few correlations found with behavioral strains (for example, Number of Cigarettes Smoked) followed predicted patterns. The infrequency of significant relationships with smoking may be due to the many influences in addition to P–E fit which affect smoking behavior. For example, if restrictions on smoking vary from job to job, they will attenuate observed relationships between P–E fit and behavior.

In contrast to the relationship of P–E fit to psychological and behavioral strains, the relationships with physiological strains were infrequent and the pattern of the relationships was inconsistent with theoretical predictions. The infrequency and weakness of these relationships is not surprising. A host of biological and environmental factors may affect the physiological strains, greatly attenuating any cross-sectional relationship between them and P–E fit. The observed relationships, however, contradicted the hypothesized relationships as often as they supported them. Of the eleven significant relationships, four were in the predicted direction, four were opposite, and three involving T3 and T4I had no predictions. Some of these relationships may have occurred by chance because most of the findings barely reached the .05 level of significance. Even allowing for the number of relationships examined and the selection from the transformations, twice as many significant relationships were observed as would be expected by chance. Further research will have to untangle the causal relationships reflected in the correlations with the physiological variables and separate out those findings which are chance from those which are replicable.

Additional Variance Accounted for by P–E Fit

The curvilinear shapes predicted by P–E fit theory were evident in the correlational analysis. The importance of the curvilinear relationships is best demonstrated by the variance in strain accounted for by the curvilinear relationship over and above the variance which is linearly associated with the component E and P measures. A series of stepwise multiple regressions were performed to determine precisely how much additional variance in strains was accounted for by curvilinear relationships with P–E fit measures. P–E fit on Income and Overtime were excluded from this set of analyses because their restricted distributions did not permit a test of curvilinear relationships between them and strains.

Each stepwise analysis began by constraining the E and P components on a P–E fit dimension to be predictors of a strain in a multiple regression equation. Then if the 'Deficiency', 'Excess', or 'Poor Fit' measure on the P–E fit dimension accounted for significant additional variance in strain, it was added to the equation. If more than one transformed P–E fit measure accounted for additional variance in strain, only the transformed measure accounting for the most variance was included in the equation. In such cases the other transformations would not account for further significant variance in strains because of the intercorrelations between transformed measures.

Assumptions and artifacts affecting results. Before the results of the stepwise multiple regression analyses are presented, several factors affecting the results and their interpretation must be considered. These factors include assumptions about the independence of the E and P components, the inherent ability of U-shaped relationships to account for more additional variance in strains than asymptotically shaped relationships, and the effect of the strength of the relationship between P–E fit and strain on the likelihood that P–E fit will account for additional variance in strains.

The statistical tests used on the nonadditive regression model (that is, ([E] + [P] + [an interaction of E and P]) assume that the E and P measures are uncorrelated. Violation of this assumption may artificially reduce or enhance the likelihood that the interaction term will account for additional variance in strains. Correlations between the commensurate E and P measures used in this analysis ranged from .13 to .75 as reported earlier in this chapter (page 34). Caution must therefore be used in interpreting the results for relationships between strains and a P–E fit dimension where the correlation was relatively high. Althauser (1971) discusses this methodological problem in detail and points out that when predictors are correlated it is more often the case that the additional variance accounted for by the interaction term will be *underestimated*. The overall pattern of the results across several P–E fit dimensions and strains will, nevertheless, provide useful information concerning the general ability of predictions derived from P–E fit theory to account for additional variance in strain. In light of these methodological considerations, the discussion of the following results focuses primarily on the overall pattern of findings.

For methodological reasons, poor fit measures should account for additional variance in strains more often than do 'Excess' or 'Deficiency' measures. This pattern of results should occur because the linear relationships of E and P measures to strains will more closely approximately the asymptotic relationships represented by the 'Excess' and 'Deficiency' measures than the U-shaped relationships represented by poor fit.

Another factor which will affect the extent to which additional variance is accounted for by curvilinear measures is the strength of the relationship between strain and the dimension of fit. The stronger the correlation between a curvilinear measure of fit and strain, the more likely that the curvilinear measure of fit will account for additional variance in strain beyond that accounted for by the component measures. When the correlations between a curvilinear measure of fit and a strain are weak, removing the variance predicted by the component measures can be sufficient to reduce the correlation of strain with the curvilinear measure of fit below the level for statistical significance.

Results of the analyses concerning additional variance. Table III.5 presents the results of the analyses that determined the specific amount of additional variance in strains accounted for by curvilinear measures of fit (Deficiency, Excess, and Poor Fit). The first column presents the multiple correlation ('R') of the E and P components with each strain. The second column indicates which of the transformed measures on the P–E fit dimension accounted for the most additional variance in the strain.[4] If none of the three measures accounted for additional variance, no entry was made. The third column presents the multiple correlations ('R') of the E, P, and selected fit measure with strain for those relationships where a transformed measure accounted for additional variance in strain. The actual amount of additional variance in a strain which is accounted for can be obtained by subtracting the square of the multiple correlation of the E and P measures from the square of the multiple correlation of E, P, and the curvilinear measures.

The results of Table III.5 shows 39 cases where either the E and P measures, alone or with a transformed fit measure, were significant predictors of strain. In 20 of these cases, curvilinear transformations on P–E fit accounted for significant variance beyond that accounted for by the E and P scores alone. P–E fit measures also accounted for significant additional variance in strain in three cases where neither the multiple

---

4. The transformed measure of fit which accounted for the most additional variance in a strain was not necessarily the one which correlated most highly with the strain in Table III.4. This shift in relationships can occur when one or both of the component measures are more highly correlated with a part of the distribution of P–E fit scores. For example, P scores on Job Complexity were more highly correlated with scores of too much (that is Excess) Job Complexity than with scores of too little (that is, Deficient) Job Complexity. Significant correlations were found between Depression and both Job Complexity-P and Job Complexity-Poor Fit. When checking for additional variance predicted by the Poor Fit measure, the variance associated with Job Complexity-P is accounted for, which also accounts for variance in strain associated with the excess scores in the Poor Fit measure. The deficiency scores have a more independent association with depression, and the result of the analysis shows that Job Complexity-Deficiency accounted for more additional variance in strain than the Poor Fit measure. Several instances of shifts of this type occurred between the correlational results and the 'additional variance' results.

correlation of the E and P measures alone nor with the additional transformed variable were significant.[5]

These overall findings support the prediction that significant amounts of variance in strains can be accounted for by the specified curvilinear relationship with fit measures in addition to the variance accounted for by linear relationships with the E and P component measures.

Turning from these overall findings, we now examine the effects of specific dimensions of P–E fit. P–E fit on job complexity did by far the best job in predicting additional variance in strain. The 14 per cent additional variance in Workload Dissatisfaction accounted for by Poor Fit on Job Complexity was the largest amount of additional variance in strain accounted for by fit on any of the dimensions. The good performance of fit on Job Complexity is to be expected given the correlational findings presented in Table III.4. Fit on Job Complexity had the greatest number of and the strongest relationships with strains. The majority of these relationships were U-shaped, indicating a sizable increase in additional variance would be found when the curvilinear measures of P–E fit were added to the component E and P measures.

At the other extreme, the poorest performance was by fit on Length of Service which accounted for no additional variance in strains. This result was anticipated from the correlations in Table III.4 which showed fit on Length of Service had only a few, weak, asymptotic relationships with strains.

One last point will be made before concluding the discussion of the results concerning the ability of P–E fit measures to account for additional variance in strain. Whenever a fit measure accounted for significant additional variance in strains, the amount of variance accounted for ranged from 1.5 per cent to 14 per cent. This range is much larger than that of 1.2 per cent to 2.7 per cent found by House (1972) in his study of job stress, or that of 1 per cent to 5 per cent found by Kulka (1976) in his study of stress among high-school students. The percentage of variance in strains accounted for by P–E fit in this study are probably due in part to the good distributions of P–E fit scores on both sides of perfect fit. In other studies the distribution of fit scores has been more restricted to a narrow range about the score of perfect fit. This narrow range has reduced the strength of relationships that could be observed in these studies.

---

5. A measure can account for a significant additional variance in a strain when the multiple correlation relating the measure and other predictors to the strain is not significant. Three instances of this nonintuitive finding occur in Table III.5. The curvilinear measure of fit and the component measures each account for some variance in strain, but not enough to be significant. When the variance in the strain accounted for by both components is controlled, the total variance in the strain is reduced, thereby increasing the proportion of variance accounted for by the 'curvilinear' fit measure. The increased proportion of variance accounted for by the 'curvilinear' fit measure may be enough for this relationship to reach the criterion for significance. When the total variance accounted for by the component measures and the 'curvilinear' fit measure is considered, however, the criterion for significance is raised since multiple predictors are likely to randomly account for more variance than a single predictor does. It is therefore possible that the total variance accounted for in the strain will not reach the criterion for significance, even though the 'curvilinear' measure accounts for significant additional variance in the strain.

Table III.5 Curvilinear P–E fit measures which account for significant variance in strains beyond the variance accounted for by the component E and P indices: multiple R

Strains	Job Complexity			Role Ambiguity			Responsibility for Persons		
	E and P	P-E fit[a]	E, P, and P-E[b]	E and P	P-E fit	E, P, and P-E[b]	E and P	P-E fit[a]	E, P, and P-E[b]
Job Dissatisfaction	.34**	Poor Fit++	.50**	.16*	Poor Fit++	.24**	.27**	Poor Fit++	.32**
Workload Dissatisfaction	.20**	Poor Fit++	.42**	.17**	—	—	.15*	Deficiency++	.22***
Boredom	.51**	Poor Fit++	.61**	.09	Poor Fit++	.20**	.30**	Poor Fit++	.37**
Depression	.12	Deficiency++	.23**	.18**	—	—	.17*	—	—
Anxiety	.07	Deficiency++	.26***	.16*	—	—	.12	Excess+	.17*
Irritation	.11	Deficiency++	.21**	.13	—	—	.10	—	—
Somatic Complaints	.13	Excess++	.21**	.11	—	—	.13	—	—
Number of Cigarettes Smoked	.11	Excess++	.22	.13	—	—	.14	—	—
Coffee and Tea	.06	Poor Fit++	.14	.01	—	—	.05	—	—
Obesity	.09	—	—	.08	—	—	.09	—	—
Heart Rate (A)	.12	—	—	.04	—	—	.04	—	—
Systolic Blood Pressure (A)	.11	—	—	.14*	Excess+	.18**	.04	—	—
Diastolic Blood Pressure (A)	.11	—	—	.13*	—	—	.07	—	—
Cholesterol (A)	.03	—	—	.06	—	—	.10	—	—
T3	.12	Excess++	.21**	.09	—	—	.07	Poor Fit+	.13
T4I	.10	—	—	.17**	—	—	.04	—	—
Serum Uric Acid (A)	.02	—	—	.09	—	—	.03	—	—
Cortisol (A)	.02	—	—	.06	—	—	.07	—	—

Table III.5 (continued)

Strains	Workload			Length of Service			Education		
	E and P	P–E fit[a]	E, P, and P-E[b]	E and P	P–E fit[a]	E, P, and P-E[b]	E and P	P–E fit[a]	E, P, and P-E[b]
Job Dissatisfaction	.27**	Poor Fit[+]	.30**	.23**	—	—	.34**	—	—
Workload Dissatisfaction	.52**	Deficiency[++]	.54**	.12	—	—	.16*	—	—
Boredom	.32**	—	—	.52**	—	—	.55***	—	—
Depression	.31**	Poor Fit[+]	.34**	.11	—	—	.17**	Poor Fit[++]	.22**
Anxiety	.14*	—	—	.05	—	—	.04	—	—
Irritation	.27**	—	—	.07	—	—	.06	—	—
Somatic Complaints	.06	—	—	.11	—	—	.23**	—	—
Number of Cigarettes Smoked	.14	—	—	.11	—	—	.29**	—	—
Coffee and Tea	.05	—	—	.04	—	—	.09	—	—
Obesity	.12	—	—	.10	—	—	.13	—	—
Heart Rate (A)	.06	—	—	.10	—	—	.09	—	—
Systolic Blood Pressure (A)	.11	—	—	.10	—	—	.21**	—	—
Diastolic Blood Pressure (A)	.01	—	—	.08	—	—	.19**	—	—
Cholesterol (A)	.07	—	—	.06	—	—	.05	—	—
T3	.11	—	—	.05	—	—	.28**	—	—
T4I	.11	—	—	.11	—	—	.10	Deficiency[+]	.16*
Serum Uric Acid (A)	.03	—	—	.12	—	—	.12	—	—
Cortisol (A)	.04	—	—	.14	—	—	.01	—	—

Note The physiological sample (n = 390) was used to determine multiple correlations involving the eight physiological measures. The random stratified sample (n = 318) was used to determine the other multiple correlations.

Specific relationships must be interpreted with caution because the E and P measures are correlated (see Althauser, 1971).

[a]The Deficiency, Excess, and Poor Fit measures were analyzed using stepwise multiple regression to determine which one (if any) accounted for the most significant variance in the strain beyond the variance accounted for by the E and P components.

[b]Uses the form of fit accounting for the most additional variance.

* p < .05 that the multiple correlation was a random finding.

** p < .01 that the multiple correlation was a random finding.

[+] p < .05 that the additional variance was a random finding.

[++] p < .01 that the additional variance was a random finding.

The amount of additional variance in strain accounted for in the present study is also appreciable when compared to the amount of variance accounted for by the components. The average amount of significant additional variance accounted for in the 20 significant relationships is 4.4 per cent (median = 3.1 per cent). On the average, this variance represents 51 per cent (median = 54 per cent) of the total amount of variance in the strain accounted for by the E, P, and fit measures together. Across all relationships where P–E fit dimensions account for additional variance, taking into account the curvilinear relationship between P–E fit and strains typically doubled the explained variance in strains.

P–E Fit and Strain: Cause or Coincidence?

The two preceding sets of analyses have demonstrated that P–E fit measures have the predicted shapes of relationships with strains and that the curvilinear relationships can account for variance in strains which is not accounted for by linear relationships with the component E and P measures. The demonstration of an association between P–E fit and strain, however, is not proof of causal relationships between them. The general theoretical model outlined in Chapter I holds that P–E fit is the immediate precursor of strains. In this section, we examine evidence concerning whether or not the observed relationships between P–E fit and strain are likely to reflect a causal sequence.

No longitudinal or experimental data are included in this study to provide direct evidence for causal effects of fit on strain. The best evidence for causality in this set of cross-sectional data is a demonstration that fit dimensions have associations with strains which are independent of other likely causal factors. The method used to check the independence of the relationships between strains and their predictors is described below. The results of the analysis are then presented and discussed.

The analysis to determine independent predictors of strains. A stepwise multiple regression was performed on each strain with almost all of the P–E fit, subjective environment, and personality measures in the study included as possible predictors. The 58 variables included in the analysis are listed at the front of Appendix D (page 142). Age and occupation were omitted. Age was known to affect some physiological strains, and they have been adjusted to remove variance due to age. The correlations between age and the other strains was known from Caplan *et al.* (1980) to be generally nonsignificant. Occupation was omitted because the measures of environment, personality, and fit were designed to account for the effects of occupation on strains (see Chapter VI).

The computer program used for the stepwise multiple regression analysis includes only cases where there were no missing data on any of the predictors in the analyses. To maximize the number of cases used to generate the final results, and thus make them more stable, the regression procedure was performed in three phases. In the first phase all of the personality, environment, and P–E fit measures (except age and occupation) were included in a forward stepwise multiple regression on each strain. Independent predictor variables were selected that met a $p < .10$ significance criterion for inclusion and that never exceeded a $p > .15$ significance criterion for removal. This procedure produced twelve or fewer likely independent predictors for each strain. Approximately two-thirds of the cases in the random stratified sample and the physiological sample were used in the first phase of the analysis.

In the second phase of the analysis, another set of forward stepwise regressions were performed, using for each strain only those predictors selected in the first phase and raising the significance criterion for inclusion to $p < .05$ and the criterion for exclusion to $p > .10$. This procedure produced ten or fewer likely independent predictors for each strain. The number of cases used in this second phase was increased by 20 per cent or more above the number of cases used in the first phase of the analysis.

In the last phase of the analysis, a set of simple multiple regressions were performed using the set of predictors for each strain obtained from the second phase of the analysis. The number of cases used in this analysis increased again because predictors not selected in the second phase were omitted. If any predictor failed to reach the .05 level of significance, the multiple regression was performed again, omitting the nonsignificant predictor. The end result of these procedures was a set of significant independent predictors of each strain using the maximum number of cases in the random stratified sample.

The results of the above procedure have an important limitation: as the number of potential predictors increases and when the predictors are inter-correlated, the likelihood of unstable or 'chance' findings increases. To assess this limitation, the instability of the findings for the psychological and behavioral strains was checked by repeating the analysis procedure described above on a second random stratified sample. The two random stratified samples had no overlapping cases. The results of this procedure generated a second list of independent predictors for the psychological and behavioral strains. A cross-validation of the two sets of predictors for a strain was performed by entering the set of predictors generated in the first random stratified sample in a multiple correlation with the strain in the second random stratified sample. Similarly, the set of predictors generated in the second sample was entered in a multiple correlation with the strain in the first sample. Predictors which accounted for significant independent variance in a strain in both samples and in the cross-validation in each sample were identified as reliable independent predictors of strain. The cross-validation could not be performed for the physiological strains because the sample for which physiological measures were available was not sufficiently large for two independent replications.

The predictors of psychological strains. The results of the analyses identifying independent predictors of strain are presented in Appendix D (page 142) and are summarized in Tables III.6 and III.7.

Table III.6 Significant independent predictors of psychological strains

Strain	Predictor
Job Dissatisfaction	Job Complexity–Poor Fit
	Underutilization
Workload Dissatisfaction	Workload–Excess
	Overtime Fit
	Workload, Quinn
Boredom	Job Complexity–Poor Fit
	Underutilization
Depression	Support from Others at Work (–)[a]
	Job Future Ambiguity
Anxiety	Deny Bad Self (–)
Irritation	Deny Bad Self (–)
	Role Conflict
	Support from Others at Work (–)
Somatic Complaints	Deny Bad Self (–)
	Education-P (–)
Number of Cigarettes Smoked	None
Coffee and Tea	None
Obesity	None

Note This table summarizes the information presented in Appendix D, Tables D.1 to D.10. All of the potential predictors included in the stepwise multiple regression analyses are presented at the beginning of Appendix D.

To be included in this table, a predictor had to have a significant independent relationship with the strain in both random stratified samples and in the cross-validation associating a strain in one sample with the list of predictors generated in the other sample.

[a]Parentheses enclosing a minus sign (–) indicate that the predictor and the strain were negatively correlated.

Table III.6 presents the independent predictors of psychological and behavioral strains. It shows that P–E fit measures were almost the only consistent predictors of Job Dissatisfaction, Workload Dissatisfaction and Boredom with significant independent relationships in both samples and the cross-validations. The direct measures of P–E fit which were independent predictors of these strains include P–E fit on the dimensions of Job Complexity, Workload and Overtime. One of the other best predictors also assesses fit, but does not use the discrepancy format. 'Underutilization' is a measure of perceived P–E fit. Instead of asking for separate ratings of the environment and person which are subsequently compared during the analysis, this measure asks the extent to which the individual's abilities (P) are utilized on the job (E). A low underutilization indicates good fit and high under-utilization indicates poor fit. The only independent predictors of any of the first three strains which are not measures of P–E fit are Future Ambiguity and a measure of workload level (Workload, Quinn). Future Ambiguity was independently related to Job Dissatisfaction. Workload level was related to Workload Dissatisfaction independently of the curvilinear relationship with the P–E fit measure, Workload Excess.

Table III.7 Significant independent predictors of physiological strains

Strain	Predictors
Heart Rate (A)	Job Complexity Poor Fit* (–)[a]
Systolic Blood Pressure (A)	Number Supervised*** (–)
	Workload-E*
	Support from Others at Work*
	Role Ambiguity Poor Fit* (–)
Diastolic Blood Pressure (A)	Education-E***
	Role Ambiguity Excess*** (–)
Cholesterol (A)	Type A Personality*
T3	Education-P***
	Job Complexity Poor Fit*
T4I	Job Insecurity**
	Service Fit**
Serum Uric Acid (A)	Service-P**
	Job Insecurity* (–)
Cortisol (A)	Overtime-E***
	Role Conflict** (–)
	Concentration** (–)

Note This table summarizes the information presented in Appendix D, Tables D.11 to D.18. All of the potential predictors included in the stepwise multiple regression analyses are presented at the beginning of Appendix D.

To be included in this table, a predictor had to have a significant independent relationship with the strain in the physiological sample.

[a]Parentheses enclosing a minus sign (–) indicate that predictor and strain were negatively correlated.

*$p < .05$
**$p < .01$
***$p < .001$

No fit measures were among the predictors of depression, anxiety, irritation, and somatic complaints. Instead, low social support, low denial of socially unacceptable characteristics about oneself, and measures of the environment had the largest consistent impact on these strains.

The pattern of results between predictors and psychological strains shows that the P–E fit dimensions predicted the strains most closely related to the job, that is, Job Satisfaction, Workload satisfaction, and Boredom. The correlations presented in Table III.4 and the multiple regressions relating P, E, and P–E fit measures to strains in Table III.5 both demonstrate that P–E fit measures were associated with the more general states of strain. However, P–E fit dimensions did not have appreciable independent effects on the more general states of strain, that is, Depression, Anxiety, and Irritation. Any effect of P–E fit on more general affective states may operate through the effect of P–E fit on the job-related strains. This possibility is examined in the next chapter.

No consistently significant independent predictors were found for the behavioral strains. However, there was an interesting tendency for Assert Good Self to be negatively related to Number of Cigarettes Smoked in the two random samples and their cross-validations ($r = -.20, -.24, -.19, -.15$). As noted earlier, this lack of findings must in part reflect the variation in behavior produced by external rules, such as restrictions on smoking and access to caffeinated beverages, in particular occupations. Such factors weaken the relationship observed between the measures of stress and the behavioral strains.

The predictors of physiological strains. The independent predictors of physiological strains are presented in Table III.7. As has been noted, no second sample was available to replicate the analysis and to cross-validate the results of the findings. The significance of the independent relationships between the predictor and the strain is included in Table III.7 to provide some information concerning the reliability of each relationship. The partial correlations in Appendix D (page 152–154) indicate that the relationships with the physiological strains were weaker than those with the psychological strains. These weaker relationships were expected because many factors, in addition to psychological stress, affect the physiological strains. Measures of P–E fit were independent predictors of five of the eight strains.

The pattern of positive and negative relationships found earlier in correlations between P–E measures and physiological strains reappears in the present analysis. For three relationships strains decreased with increasing misfit, a finding opposite to the expected direction. Various P–E fit dimensions may differentially affect the physiological strains or perhaps one P–E fit dimension may differentially affect various physiological strains. The general assumption in selecting this set of strains, however, was based on previous research where one or more of them showed increases with the occurrence of psychological, occupational stress. The observed relationships support continued use of P–E fit measures to predict physiological strains, while casting doubt on the general prediction that physiological strains will only increase with increases in misfit. These findings also cast doubt on the concept of *generalized physical strain*— in other words, the expectation that all physiological responses change in a given direction in response to *all* stresses.

Is misfit the cause of strain? Two important questions must be addressed in interpreting the results of the analyses concerning significant independent predictors of strain. One question concerns the strains for which no P–E fit measure was found to be a significant independent predictor. Do these findings indicate that the P–E fit dimensions have no causal effect on the strain? The second question concerns the strains for which P–E fit measures were found to be significant independent predictors. Do these findings indicate that the misfit causes the strain?

The failure of a variable to be a significant independent predictor of a strain does not rule out the possibility that the variable has a causal effect on strain. For example, assume that Workload causes role conflict, which, in turn, causes irritation with co-workers and superiors. Assume this is the only path for linking Workload and Irritation, so that role conflict has the highest correlation with Irritation. If both Workload and Role Conflict are entered in a stepwise multiple regression as predictors of Irritation, only Role Conflict will emerge as a significant independent predictor. The intervening variable, Role Conflict, will be selected because its relationship with Irritation will be stronger than that between Irritation and the antecedent condition of Workload. Removing the variance in Irritation associated with Role Conflict will leave little variance which is independently associated with Workload. Therefore, only Role Conflict will appear as a significant independent predictor of Irritation even though both Workload and Role Conflict are related to Irritation in a causal chain. Variables with a true causal relationship to strain may be omitted from the results of the stepwise selection procedure because of other underlying relationships between predictors. Predictors may be correlated with each other because they measure overlapping constructs (for example, Underutilization and Job Complexity-Deficiency) or because they tend to co-vary 'naturally' (for example, Job Complexity-E and Income E). In such instances the inclusion of the variable with the stronger relationship to strain is likely to account for most of the statistical association between the other variable and strain. It must be assumed that selecting variables having the strongest relationships to strains identifies predictors that are likely to have the most *immediate* and *direct causal relationship* to strains. The predictors excluded from Table III.6 and Table III.7 may still have causal effects on strains, but their association is empirically weaker and likely to be less immediate or direct than those of the significant independent predictors identified by the analysis.

Does a significant independent relationship between P–E fit and a strain demonstrate that misfit causes the strain? Before concluding that a causal relationship exists between any of the independent predictors and strains, it must be recalled that these results are open to other interpretations. It can be argued that the relationship results from association with an unmeasured variable which is the true cause of the strain. This argument cannot be disproved because no evidence has been obtained concerning the unmeasured variable.

An alternative interpretation is that the true causal direction is from the strain to the level of P–E fit. It is hard (but not impossible) to see how the level of an asymptomatic physiological variable affects a predictor. It is easier to suggest reverse causal relationships between the psychological strains and their predictors. One could alter one's perceptions of self and the job to bring them in line with one's feelings about the job. For example, a worker who is dissatisfied with the job may start to perceive either the complexity of the job (E) or his or her abilities (P) differently so as to increase misfit on Job Complexity.

The two longitudinal studies of fit and strain that we know of suggest that fit has a causal effect on strain. Vickers (1979) finds that misfit on a dimension at one point in time can account for part of the change in a strain between that point and a later point in time. His results suggest that individuals experiencing misfit reduce strain by subsequently either using cognitive defense mechanisms to improve subjective fit or altering the objective levels of E and P to improve objective fit. Campbell (1974) also finds that P–E fit measures are more likely to affect the levels of strains across time than vice versa. The results of these longitudinal studies and the pattern of relationships in the present cross-sectional study all support the hypothesis that misfit can cause strain.

Before concluding the discussion of the findings in Table III.6 and Table III.7, it is worthwhile to note that five component measures of fit variables were also significant independent predictors of physiological strains: Workload-E, Length of Service-P, Education-E, and Education-P. These results serve as a reminder that P–E fit variables supplement, not replace, the predictive power of component measures.

Summary

Person–environment fit theory holds that strain can result from the mismatch between the person and the environment on dimensions important to the wellbeing of the individual. The point of perfect fit between the person and the environment is hypothesized to be a turning point for U-shaped, asymptotic, and linear relationships likely to exist between measures of P–E fit dimensions and strains. These curvilinear relationships were found to account for significant additional variance in strains in over half of all relationships where the strain had a significant linear relationship with a P–E fit measure or with a component measure of the environment or the person. The amount of significant additional variance typically doubled the total amount of predicted variance in strain.

Causal relationships between P–E fit and strains were suggested by analyses which found that of the several dozen independent variables measured in the study, five P–E fit dimensions had significant independent relationships with one or more of eight strains. The predicted relationships were strongest for psychological measures of strain. As expected, the relationships with physiological measures of strain were fewer and weaker. Unexpectedly, the relationships with physiological measures were the inverse of expected shapes as frequently as they followed the expected shapes. Further studies focusing more directly on physiological measures will be required to unravel these unexpected findings.

The findings supporting P–E fit theory emphasize that job stress must be understood in light of the relationship between the job and the individual. This interaction between the job and the person emphasizes the importance of personnel section when hiring and transfering employees and the equal importance of allowing individualization of the job to fit the needs and values of each worker. These and other implications of P–E fit theory for reducing job stress are considered in more detail in the final chapter.

Chapter IV

Job Demands, Strain, and Intervening Variables

The preceding chapter discussed the theory of person–environment fit and tested predictions of the theory. The analyses focused on the relative importance of P–E fit versus its components, E and P as predictors of strain. As part of those analyses, we reported findings on the extent to which P–E fit showed linear and nonlinear relationships with strain.

In this chapter, the focus shifts. Here we examine the likely causal paths linking P–E fit, job demands, job-related affects, general affective strains, and physiological strains. Within the obvious limitations of a cross-sectional set of data, an attempt is made to suggest which of these predictors are antecedent and which are intervening variables. For example, analyses are performed which suggest that job-related affects intervene between job demands and general affective strains.

The chapter examines the total amount of variance in hypothesized intervening and dependent variables that can be explained by predictors that have significant main effects on these variables. Data are presented that show the relative contributions of each of the predictors. Thus, if several types of job demands predict to Job Dissatisfaction, the analyses show the total effect of all of these demands as well as their relative contributions controlling for the effects of the other job demands.

Univariate Analyses Suggest Sequential Model

In Caplan *et al.* (1980) we presented the correlations among the various measures of stress, of the person, and of strains. The resulting inter-correlations have been abstracted into a correlogram which appears here as Figure IV.1. The figure includes correlation coefficients between stress and strain which were $\geq .30$. For correlations among the strains we have included only those coefficients $\geq .20$. Although Job Complexity–E and Responsibility for Persons–E had relationships meeting these requirements, they were omitted from Figure IV.1 because they are hypothesized to affect strain through P–E fit on these dimensions and because the P–E fit variables on these dimensions had higher correlations with strains.

59

60

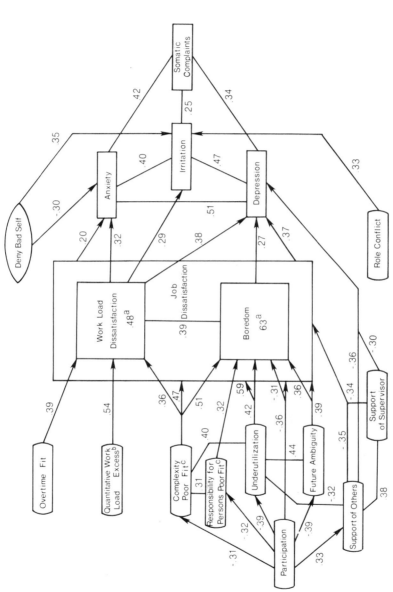

Figure IV.1   An interpretation of major correlations between stresses and psychological strains at the individual level

Note All correlations are based on approximately 310 men from the random stratified sample. Stress–stress and stress–strain correlations ≥ .30 and strain–strain correlations ≥ .20 are presented (except for the Job Complexity-E and Responsibility for Persons-E indices — see text). Arrows indicate suspected causal relationships; lines without arrowheads are interpreted as non-causal. Where more than one form of fit was used, the form that accounted for the most variance was selected as the predictor.

A Sequential Model

The correlogram shows that many of the measures of job demands were positively correlated. The same can be observed for many of the measures of strain. Beyond such patterns, the relationships among the different classes of variables suggest a basic model that is summarized in Figure IV.2. The model suggests that specific job demands might produce job-related strains such as job dissatisfaction and boredom. These job-related strains might then produce more general states of strain such as anxiety, depression, and somatic complaints.

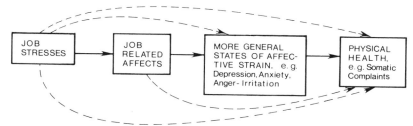

Figure IV.2 A model of relationships among stress, affects, and physical health. A dashed arrow represents a derived relationship; for example, if the three solid arrows represent true causal relations, then job stresses will have an indirect effect on (and will be correlated with) more general strains

The correlogram in Figure IV.1 also suggests that participation may be antecedent to good person–environment fit (Campbell, 1974) and that good fit, in turn, may lead to low levels of job-related strain such as boredom. In this chapter we test the plausibility of this sequential model.

Multivariate Results: Predicting Affects, Somatic Complaints, Physiological Strain, and Illness

Methods of Analysis

The analyses deal with three categories of variables — independent, intervening, and dependent. In order to test the relationships between hypothesized independent, intervening and dependent variables, we conducted two forms of analyses, as follows.

Analyses of the relative contributions of a variety of predictors. These contributions were determined using stepwise multiple regression. This method was used, for example, when examining the relative contributions of all of the significant predictors of Job Dissatisfaction. It was also used (a) to examine the multivariate effects of the job-related strains on the general affects and on Somatic Complaints, and (b) to examine the multivariate relationships of the job-related and general affects on Somatic Complaints.

Analyses of the relative contributions of hypothesized intervening variables. These analyses determined (a) the extent to which the job-related affects intervened between job demands and the general affects and Somatic Complaints, and (b) the extent to which the general affects intervened between the job-related affects and the Somatic Complaints. In planning these analyses, we considered two methods.

First, one can remove statistically the effect of the intervening variables on the dependent variable and recompute the effects of the predictor on the residuals. This method is not preferred because it introduces a downward bias in the resultant coefficient in proportion to the amount of correlation between the predictor and the intervening variables (Andrews *et al.*, 1973; Goldberger and Jochem, 1961; Morgan, 1958).

Alternatively, one can remove the effects of the intervening variable from the predictor in question, and then correlate the residuals of that predictor with the dependent variable (Andrews *et al.*, 1973). The result is the part correlation. The part correlation, when squared, provides an estimate of the marginal or unique effects of the predictor relative to the total amount of variance in the dependent variable. This information can be obtained by running two multiple regression analyses and using the following formula adopted from Formula 4.25 of Andrews *et al.* (1973).

$$\text{Squared part correlation} =$$
$$(R^2 \text{ with the inclusion of all predictors}) - (R^2 \text{ omitting the predictor of interest})$$

For the purposes of this chapter, we selected predictors that correlated .30 or higher with hypothesized strains. We also attempted to select predictors that were not highly intercorrelated with one another in order to meet the assumption of independence of predictors for multiple regression analyses. This criterion is particularly germane when selecting among the P, E, and P–E fit measures of a job-related dimension (such as role ambiguity) where the P and E indices are components of the measure. We selected the most predictive variable of the three—usually the P–E fit measure. Among the several forms of P–E fit scores, we selected the most predictive one based on the analyses reported in Chapter III.

The presence of strongly correlated and conceptually similar measures in the stepwise multiple regressions in Chapter III does lead to the risk that one of the predictors will be selected eliminating the other, not because the other is weak, but because it is highly related. Such effects are less likely in the analyses performed in this chapter because only the strongest version of each predictor was selected.

These differences in the way that predictors were selected for the purposes of this chapter and for the purposes of Chapter III can lead to somewhat different results when one conducts the multiple regression analyses to examine the relative contributions of the predictors of the strain. For the most part, however, the pattern of results is quite similar. In Chapter VII we compare the two sets of findings demonstrating that this is the case.

Thus, this chapter has focused on the total and relative contributions of the most important predictors of strain, whereas Chapter III examined the relative contributions of all predictors regardless of their initial statistical significance as main effects. The latter strategy might be advisable if there are suppressor effects (Rosenberg, 1968), but we have not uncovered any such effects of significance in these analyses.

In summary, procedures followed in Chapter III are appropriate for determining whether P, E, or P–E is the best predictor of strain. The methods of analysis used in this chapter, however, are most appropriate for determining (a) the relative and total contributions of relatively independent facets of job-related stress, and (b) the role of intervening variables.

Predictors of the Job-Related Affects

This section describes significant predictors of the job-related affects.

Dissatisfaction with Workload. This strain had 45 per cent ($R = .67$, $p < .001$) of its variance accounted for by three measures of job demands. These were Workload Excess (partial $r = .51$, $p < .001$) and Job Complexity Poor Fit (partial $r = .34$, $p = < .001$), and Unwanted Overtime (partial $r = .29$, $p < .001$).

Boredom. Forty-five per cent of the variance in Boredom ($R = .67$, $p < .001$) was accounted for by its significant predictors. The partial correlations for the predictors were as follows: Underutilization, partial $r = .48$, $p < .001$; Complexity Poor Fit, partial $r = .32$, $p < .001$; Responsibility Poor Fit, partial $r = .13$, $p = .02$.

Two other predictors, Job Future Ambiguity (partial $r = .10$, n.s.) and Participation (partial $r = -.03$, n.s.), failed to enter the regression equation after the previously reported variables had been entered. As noted earlier in this chapter, high participation was associated with high levels of Utilization and person–environment fit (the correlations ranged between .31 and .39). The results of the stepwise multiple regression reported here are further evidence that participation may have its effects on boredom primarily via improved person–environment fit because participation accounted for no additional variance in boredom after these variables were considered. This set of relationships with participation is summarized in Figure IV.3.

Job Dissatisfaction. Thirty per cent ($R = .55$, $p < .001$) of the variance in this strain was accounted for by its correlated measures of job demands and fit. The partial correlations for the predictors were: Underutilization, partial $r = .24$, $p < .001$; Participation, partial $r = -.15$, $p = .007$; Job Future Ambiguity, partial $r = .16$, $p = .006$; Social Support from Others at Work, partial $r = -.18$, $p = .001$; and Job Complexity Poor Fit, partial $r = .06$, *ns.*

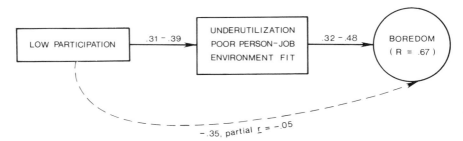

Figure IV.3 Summary of first-order correlations among Participation, Person–Environment Fit and Boredom. Arrows represent hypothesized directions of effect. The dashed arrow indicates a derived relationship

Although 46 per cent of the variance in Job Dissatisfaction (R = .68, p < .001) was accounted for by Boredom and Workload Dissatisfaction (respective partial rs = .54 and .33, both p < .001), the analyses showed that the most important predictors of each of these three strains tended to be somewhat different. Unwanted Overtime and Workload Fit were predictors of Dissatisfaction with Workload but were not predictors in the stepwise multiple regression of the other two job-related affects. Responsibility Fit and Underutilization were predictors of Boredom but not of Workload Dissatisfaction nor of Job Dissatisfaction. Job Complexity Poor Fit was correlated with Workload Dissatisfaction and Boredom but not with Job Dissatisfaction. And Participation was correlated only with Job Dissatisfaction.

Implications of these findings. The pattern of the results supports the *relevance hypothesis* that the strongest relationships between stress and strain will occur when both are measured on commensurate dimensions. Thus dissatisfaction that specifically dealt with workload was best predicted by measures of quantitative workload such as reported Overtime, Workload Excess, and to a lesser extent by a measure of qualitative workload, Complexity Poor Fit. Boredom, on the other hand, was best predicted by measures of the qualitative nature of the workload—Complexity Poor Fit, Responsibility for Persons, and Underutilization of Skills and Abilities, but measures of quantitative workload were not important predictors. Consistent with these patterns of findings, it was not surprising to find that Job Future Ambiguity was unrelated to either Boredom or Dissatisfaction with Workload.

The findings suggest that although employees do have job demands that are sources of general dissatisfaction with the job, there are job demands that may have much stronger effects on more specific job-related affects. Some of these job demands may have effects on strain which go unnoticed if one only considers general, nonspecific measures of job dissatisfaction.

The principle of relevance may suggest *new* measures of strain that have not been examined in this study. That principle, as noted in

Chapter I, states that the strain most likely to be affected by a stress is the strain which is most commensurate with it. For example, if we are studying the job demand of Responsibility for Others, we may be able to increase the relationship of this demand to strain if we create a measure of strain specifically designed to tap its effects. One such measure of strain might be the extent to which the employee is satisfied with the amount of responsibility for other people. Although general job dissatisfaction may include such satisfaction, we should expect that the association between responsibility for persons and such a global measure will be watered down by the effects of other types of job demands unrelated to responsibility for persons.

Predictors of the General Affects and Somatic Complaints

These findings are summarized in Table IV.1. For each general affect (Anxiety, Irritation, and Depression) and Somatic Complaints, the following information is presented.

1. The predictors which showed significant bivariate correlations are presented in the left-hand column. These predictors are generally measures of subjective environment and of subjective P–E fit. Following each predictor is its partial correlation with the affective or somatic strain. For example, in the first line of data in the table, Unwanted Overtime has a partial correlation of .18 with Anxiety.
2. In the middle column, we have listed strains which are hypothesized to intervene between the predictors and dependent strains according to the correlogram in Figure IV.1 and the derived Figure IV.2. The correlation of each intervening strain with its dependent strain, after partially out the effects of the other strains intervening between the predictors and dependent strains, is presented after each intervening strain. Referring to the first line of data in the table as an example, Workload Dissatisfaction had a partial correlation of .26 with Anxiety when Boredom and Job Dissatisfaction were controlled.
3. The right-hand section lists three squared multiple correlation coefficients, R_1^2, R_2^2, and R_3^2. R_1^2 is the percentage of variance in the predicted strain accounted for by the variables in the left-hand column. The variables in the left-hand column are usually measures of subjective environment and subjective person environment fit. R_2^2 represents the percentage of variance accounted for by the variables in the middle column, the intervening strains. R_3^2 is the percentage of variance accounted for by the variables in the left-hand and middle columns combined.

One can get an estimate of the unique contribution of variance of all the variables in either column to the predicted strain by subtracting the R^2 for the respective column from R_3^2.

Now the findings for each general affective and somatic strain are discussed.

Table IV.1 Percentages of variance in strain accounted for by multiple predictors

Predictors	Partial r of predictors controlling other dependent strains	Intervening strains	Partial r of intervening strains with dependent strains	Dependent strains	R_1^{2a}	R_2^2	R_3^2
A. Unwanted Overtime	.18^c	Workload Dissatisfaction	.26^d	Anxiety	.11^d	.11^d	.15^d
Complexity Poor Fit	.14^b	Boredom	-.05				
Support from Others	-.20^d	Job Dissatisfaction	.08				
B. Unwanted Overtime	.12	Workload Dissatisfaction	.26^d	Irritation	.14^d	.09^d	.16^d
Workload Excess	.25^d	Boredom	-.02				
Support of Supervisor	-.03	Job Dissatisfaction	.01				
Support from Others	-.16^b						
C. Workload Excess	.23^d	Workload Dissatisfaction	.03	Depression	.21^d	.19^d	.26^d
Complexity Poor Fit	.07	Boredom	.24^d				
Participation	.00	Job Dissatisfaction	.18^c				
Underutilization	-.05						
Future Ambiguity	.11						
Support from Supervisor	-.15						
Support from Others	-.21^d						
D. Workload Dissatisfaction	.10	Anxiety	.30^d	Somatic complaints	.04^d	.20^d	.20^d
Job Dissatisfaction	.10	Depression	.14^b				
		Irritation	.04				
E. Complexity Poor Fit	.08	Anxiety	.30^d	Somatic complaints	.05^d	.20^d	.20^d
Underutilization	.09	Depression	.12^b				
Support from Others	-.11	Irritation	.05				
		Job Dissatisfaction	.03				
		Boredom	.03				
		Workload Dissatisfaction	-.04				

Note Only predictors which showed significant bivariate correlations with the dependent strains have been included as the stresses and intervening strains in these analyses. N ranges between 270 and 313, depending on the amount of missing data. The letters in the left-hand column ('A', 'B', 'C' and so on) identify sets of predictors for one dependent strain. For example, 'A' marks off all the predictors of anxiety.

$^a R_1^2$ = per cent of variance in the dependent strains accounted for by the environmental variables in the left-hand column. R_2^2 = per cent of variance in the dependent strains accounted for by the intervening strains in the middle column. R_3^2 = per cent of variance in the dependent strains accounted for by the variables in the left-hand and middle columns combined.

$^b p < .05$ $^c p < .01$ $^d p < .001$

Anxiety. These results are presented in the rows of section A of Table IV.1. Low Social Support from Others, Poor Fit on Job Complexity, and Unwanted Overtime were the main stresses predicting Anxiety. Workload Dissatisfaction, but not Boredom or Job Dissatisfaction, was the main intervening job-related strain that predicted to Anxiety. R^2 increased from .11 to .15 when the measures of stress were added to the intervening predictors. Although most of the variance in stress had its effect on anxiety via the intervening strains, stress and strain each independently explained 4 per cent of the variance in Anxiety.

Irritation. The main stresses related to Irritation, according to the multiple regression, were Workload Excess and Social Support from Others. Workload Dissatisfaction was the main intervening job-related strain predicting Irritation. A comparison of the R^2 values in the right-hand columns shows that 7 per cent of the variance in Irritation was accounted for by the measures of stress, independent of the effects of the intervening strains. Most of the variance in Irritation that was due to the main effects of the intervening strains appeared to be due to the hypothesized antecedent measures of stress.

Depression. Depression varied as a function of the independent effects of the measures of stress and of the measures of the intervening strains. Among the measures of stress the indices of Workload Excess and Social Support from Others were the main predictors of depression. Both Boredom and Job Dissatisfaction had independent effects on Depression.

When the measures of stress were added to the intervening strains, they accounted for 7 per cent additional variance in Depression. As can be seen, the R_3^2 value was also higher by five percentage points than the R^2 for just the measures of stress (.26 versus .21 respectively). Consequently, both the measures of stress and the intervening strains added some unique predictive power in explaining variance in Depression.

Somatic Complaints. We had two links to explore here. The first was the relationship among the job-related affects, the general affects, and Somatic Complaints as three separate sets of variables. The correlogram in Figure IV.1 suggested that the general affects intervene in the relationship between the job-related affects and the Somatic Complaints. The relevant results are presented in the rows of section D of Table IV.1.

The two job-related affects that predicted to Somatic Complaints were Workload Dissatisfaction and Job Dissatisfaction. They had partial correlations with the complaints measure that approached statistical significance ($p < .10$). Of the general affects, Anxiety and Depression, but not Irritation, had significant partial correlations with Somatic Complaints. The pattern of R^2 coefficients in the right-hand columns clearly suggests that although the job-related affects did affect variance in Somatic Complaints, all of these effects (4 per cent of the variance) operated via the general affects.

The second link explored dealt with the relationship between job stress and Somatic Complaints (the rows in section E). This analysis included the job-related and general affects as the intervening variables. None of the measures of stress had significant partial correlations although Underutilization was significant at $p = .09$ and Social Support from others was significant at $p = .06$. The pattern of R^2 values showed that all of the effect of the measures of stress on Somatic Complaints (5 per cent of the variance) was accounted for by the intervening strains.

Discussion and summary. Figure IV.4 summarizes the findings from these analyses. It presents a model in which job demands or stresses produce job-related affects. The job-related affects, in turn, produce general affective states such as Anxiety and Depression. In turn these states may produce

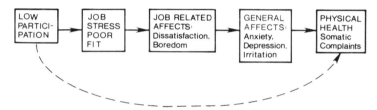

Figure IV.4 Summary of analyses using multiple regression and partial correlation. Dashed line indicates derived relationships expected if the solid arrows represent true causal relationships

somatic reactions such an inability to fall asleep easily, trembling hands, dizziness, and loss of appetite (or, at least an increased tendency to report such symptoms). This pattern has also been found by Vickers (1979) in a study of white-collar employees at NASA. One must keep in mind that Figure IV.4 omits the conclusion that there must be other unknown mechanisms by which job demands determine general affects and somatic complaints. As noted in Caplan *et al.* (1980), there were no convincing relationships between job demands and either physiological measures or reported illnesses.

With regard to the relationship of job stress to job-related affects, the analyses in the first part of this chapter suggest that the relationships depend on the stress and the affect which are specified. Although employees can make general assessments of their satisfaction with the job as a whole, their more specific assessments of satisfaction with workload and of boredom are most likely to be related to unwanted workload and poor utilization of their skills respectively. General job dissatisfaction appears less strongly related to such strains.

Participation is depicted in Figure IV.4 as having beneficial effects on person–environment fit on the job. The findings suggest that participation may be an important organizational mechanism for allowing employees to improve their adjustment to the demands of the job by having a say in the decisions which determine those demands.

Physiological Findings

In Caplan *et al.* (1980) the lack of support for the hypothesis that stress produces physiological strain was thought to be due, perhaps, to the fact that many of the physiological measures were confounded with variables such as smoking or age. For this reanalysis, consequently, we took steps to control statistically for such variables. The results were no more encouraging. We examined a total of 584 correlations, correcting for confounding variables where appropriate. About 10 per cent were significant at $p < .05$. Only a couple of the coefficients were above .20, the maximum being .22, and these correlations were associated with years of schooling rather than job stress. The other coefficients were all lower, being primarily less than .15. The pattern of the findings did not reveal any consistent trends.

These findings may suggest that psychosocial factors have little or no impact on physiological functioning. Such a conclusion, however, seems very unwarranted. Experimental studies have firmly established the significance of psychological variables on human physiology (see, for example, Levi, 1972; Mason, 1973; Obrist *et al.*, 1974).

We noted that the relationships between job demands and more general affects and Somatic Complaints were weaker than the relationships between job demands and more commensurate intervening dimensions of strain such as Job Dissatisfaction. The chain of intervening variables between job demands and physiological response is longer, and the likelihood of uncovering detectable relationships in this cross-sectional survey design may be more remote.

The Illness Findings

As detailed in Caplan *et al.* (1980), four illness categories appeared with sufficient frequency for analyses: Cardiovascular Disease, Respiratory Infection, Gastrointestinal Problems, and Peptic Ulcer. In that report we found negligible relationships between job demands and self-reported illness. One of the strongest positive associations between a psychological variable and illness was that between anxiety and illness.

In analyses for this book we decided to break the illness category, Cardiovascular Disease, into its two original components. These were coronary heart disease and high blood pressure. The two disease groups had been combined originally in order to produce a large sample of persons in the category, Cardiovascular Disease.

Tables IV.2 and IV.3 show the significant findings. Both self-reported heart disease and high blood pressure were correlated positively with age (both rs = .16, $p < .001$). We inspected the covariates of these diseases in the two tables

Table IV.2 Covariates of coronary heart disease as reason for most recent physician visit

| | Coronary heart disease | | | | | | |
| | No (N = 1935) | | Yes (N = 69) | | | | |
Covariate	X	SD	X	SD	F	eta	P
Type A	5.17	.92	5.40	.92	4.07*	.04	.04
Somatic Complaints	1.28	.30	1.35	.40	3.85	.04	.05

*When age is controlled statistically, F = 1.31, n.s., eta = .025.

Table IV.3 Covariates of high blood pressure as reason for most recent physician visit

| | High blood pressure | | | | | | |
| | No (N = 1909) | | Yes (N = 94) | | | | |
Covariate	X	SD	X	SD	F	eta	p
Anxiety	1.68	.53	1.83	.65	7.83	.06	.005
Depression	1.63	.51	1.78	.65	6.57	.06	.01
Irritation	1.80	.50	1.91	.50	4.64	.05	.03
Job Dissatisfaction	3.29	.85	3.56	.93	9.18	.07	.002
Workload Dissatisfaction	2.18	.94	2.44	1.00	6.96	.06	.008
Somatic Complaints	1.28	.29	1.35	.40	4.90	.05	.03

Note Type A did not predict to reported high blood pressure.

to see if they were also correlated with age. Type A was correlated .12 (p < .01) with age in the total sample.[6] Its association with reported visits for coronary heart disease disappeared when age was controlled statistically.

The other covariates were either uncorrelated with age or were weakly, but negatively, correlated with age (for example, Somatic Complaints was correlated -.08, p < .05; Anxiety was correlated -.10, p < .05). Controlling for these effects of age would increase the statistical significance of the findings. Attempts to make such corrections were viewed as capitalizing on chance because the direction of the correlations was counter to what might be expected; consequently no such corrections were made.

The findings show that persons who reported that their most recent visit to the physician was for either coronary heart disease or high blood pressure were more likely to have high scores on somatic complaints. Persons who reported their most recent visit was for high blood pressure also had significantly higher levels of Anxiety, Depression, Irritation, Job Dissatisfaction and Workload Dissatisfaction.

The magnitude of these relationships is very weak when one examines the etas in Tables IV.2 and IV.3. This is to be expected because the measure of

6. As noted in Chapter II, the total sample was used to analyze data involving the illnesses, whereas the stratified random sample was used for most other analyses examining only questionnaire data.

illness is a very indirect one. It is further to be expected because the link between these affects and illness is indirect, being mediated by intervening physiological responses that eventually result in illness.

The interpretation of causal direction, of course, must be a very cautious one. Discovering that one is diagnosed as having coronary heart disease or high blood pressure could lead to the affects described in the tables. Patients may become dissatisfied with their jobs in the belief that it was the job that produced the illness. And the diagnosis of illness could increase the person's attention to bodily sensations and increase somatic complaining. The cross-sectional design of this study cannot sort out the relative effects of each of these hypothesized sequences.

Note that there were no measures of either person–environment fit or job demands that were associated significantly with the reported reasons for visits to the physician. There were two nonsignificant trends. Misfit on Job Complexity was likely to be highest among persons reporting high blood pressure as the reason for recently visiting the physician ($F = 3.00$; $df = 1$, 1955; $p = .08$) and misfit on Responsibility for Presons was also likely to be highest for such persons ($F = 3.53$; $df = 1$, 1962; $p = .06$). The absence of significant effects on illness of measures of job demands and fit could be due to their remoteness from illness in our model. The model links job stress to illness via intervening affects and physiological responses. If the reported illnesses alter the employee's affective wellbeing, then the effects of perceived illness on perceived job demands would be similarly remote, being mediated by intervening general and job-related affects. If the findings in Tables IV.2 and IV.3 merely reflect the effects of knowledge of illness on affective state, then one would not expect the relationship between job demands and illness to be very strong. In such a case, illness would be the antecedent set of variables and job demands would be the consequent set.

Research by House and Sales (House, 1972; Sales and House, 1971) suggested that extrinsic motivation was positively associated with risk of coronary heart disease and that intrinsic motivation was negatively associated with risk of the disease. Our study did not include measures of these motivations, but an attempt was made to identify items that might have content representing either of these motives. In this way we could determine if they predicted to cardiovascular disease. This proved to be nonproductive.

One of the measures, however, showed a nonsignificant trend in the expected direction. This was the single item measure of how much the person *preferred* role demands requiring concentration. This measure seemed like a good indicator of intrinsic motivation. Persons preferring work where they could concentrate on the task itself would be expressing a motive representing an intrinsic aspect of the work, rather than its pay or the opportunities for prestige and promotion that it might offer. Preference for work that required concentration was associated with low Coronary Heart Disease (eta = .04, $p = .06$). Persons reporting Coronary Heart Disease had lower scores on preference for concentration than persons not reporting such illness (mean = 4.8, SD = 1.0, and mean = 5.1, SD = .9, respectively). The reader is reminded that these were illnesses which the employee reported as the reason for the most recent visit to a physician for purposes other than a routine physical examination. There was no association between preference for concentration in the work role and the reported incidence of high blood pressure as the reason for the visit to the physician.

As part of these analyses, we formed an additional index of illness that summed the total number of illnesses mentioned by each respondent. We examined the correlation between that index and measures of stress and strain to see if there was any evidence of a hypothesized positive effect of job demands on illness. Again the results were quite discouraging. The three highest correlations were between total illnesses and age ($r = .15$, $p < .001$), Anxiety ($r = .12$, $p < .001$), and Somatic Complaints ($r = .15$, $p < .001$). Although some of the other coefficients were statistically significant, they were even lower, all being less than .10. In reviewing the weak pattern of findings relating job demands and illness, one must keep in mind that the study was not designed to examine physical illness, and that we included the measures of illness as an incidental attempt to examine the effect of job stress on illness.

Chapter V

Some Further Interaction Effects
of Job Demands on Strain

In Chapter III we tested the hypothesis that the conditions of work may best predict to employee wellbeing when one knows how those conditions interact with the needs and abilities of the employee. This theory of person–environment fit was tested using commensurate dimensions of the job environment and of the person. In this chapter we examine some other forms of interaction of job and person characteristics. Some of the hypothesized moderators of the effects of job demands on strain have to do with states of the person, such as the worker's anxiety at the time a job demand is made. Other moderators have to do with additional conditions of the work itself that may alter the meaning of the job for the person. In this chapter, these moderators of the effects of job demands are measured on noncommensurate dimensions, and so they represent hypotheses that are not formally within person–environment fit theory (French, Rodgers and Cobb, 1974).

Hypotheses Regarding the Conditioning Effects of Strain, Type A Personality, and the Work Environment

H1: Conditioning Effects of Strain

The person's existing state of psychological strain may condition the effects of job demands on other psychological and physiological strains. More specifically, the *relationship* between job demands and any one strain should be the strongest for persons already under other strain. Persons already experiencing strain may have low ego resiliency (Block, 1965) which makes it difficult for them to cope effectively with job demands. Their affective and physiological responses to job demands may be stronger than persons who are not already in states of anxiety or depression.

To explore this hypothesis economically, the general affective strains (Anxiety, Depression, and Irritation) were designated as conditioning variables. The work-related affects (Boredom, Workload Dissatisfaction, and Job Dissatisfaction) and physiological variables were used as dependent variables. Thus we examined the conditioning effects of general affective strains on the relationship between job demands and both job-related affects and physiological responses.

Person–environment misfit, like the job demands, was similarly hypothesized to have stronger effects on strain when the general affective strains of the employee were high. Where the main effect of fit on strain was already known to be curvilinear (for example, Chapter III shows that strain increases as $P > E$ or as $E > P$), this curvilinear relationship was expected to be strongest under conditions of high general affective strain.

H2: Conditioning Effects of Type A Personality

Traits of the person, such as personality Type A, were predicted to condition the effects of stress on strain. The relationship between job demands and strain was expected to be stronger for Type A persons than for Type Bs. The high involvement of Type A persons in their work has been suggested as the reason why they may show strong reactions of strain to their job demands. Previous studies have found this to be the case for the relationship between quantitative work load and physiological strain (Caplan, 1972; Caplan and Jones, 1975) as well as psychological indicators of strain such as anxiety and depression (Caplan and Jones, 1975). Gurin, Veroff and Feld (1960), in *Americans View Their Mental Health*, found that persons who were highly involved in their work experienced the greatest joys but also the greatest disappointments depending on whether things went well or not at work.

Research by Glass (1977) suggests that Type A persons have a higher need to control their environment than Type B persons. As a result, Type A persons may experience more emotional and physiological strain when exposed to uncontrollable events. In this study neither need for control nor the controllability of stress was measured. We did assess, however, a potentially important mechanism for controlling job demands — participation in decision-making. Analyses in Chapter IV suggested that it was appropriate to view participation as such as mechanism. Participation tended to be highest for persons with good person–environment fit. Furthermore, the negative association between participation and job-related strain was accounted for by the positive effects of participation on person–environment fit.

In an extension of the hypothesis about high need for control among Type A persons, we hypothesized that the tendency for job demands to produce strain should be strongest among Type A persons, particularly when they experience little means of controlling the demands, as indicated by low levels of Participation. We examined blue-collar workers separately from white-collar workers on the grounds that participation might be perceived as a more legitimate mechanism by white-collar workers (French, Israel and Aas, 1960). Hence any absence of participation might be more frustrating to white-collar than to blue-collar workers when stress was present. When the analyses examined physiological dependent variables, blue- and white-collar workers were pooled because the sample size was small and a maximum sample was needed to test this triple interaction reliably.

H3: Conditioning Effects of the Organizational Environment

Certain organizational aspects of the job environment were hypothesized to buffer the effects of job demands on strain. Participation was hypothesized to condition the effect of the job environment on strain. The main effects of participation on strain were reported in Chapter IV. There, multivariate analyses suggested that participation reduced strain by improving person–environment fit and by reducing job demands. Although those analyses were cross-sectional they are supported by longitudinal findings from another study (Campbell, 1974). That study found that participation predicted to later improvements in person–environment fit but fit did not predict to subsequent participation.

Participation, as a conditioning variable, would be expected to buffer the effects of job demands on strain. Persons with high participation should perceive that if they are overloaded with job demands or if there is poor fit, they can reallocate the role demands through the mechanism of participation. Consequently, the association between job stress and strain should be strongest for persons with low reported participation in decision-making.

H4: Conditioning Effects of the Qualitative Content of the Job on the Relationship between the Quantitative Workload of the Job and Strain

Whether or not poor person–environment fit on *quantitative* workload leads to strain may depend on the extent to which a person has poor fit with regard to the *qualitative* nature of the work. Whereas the quantitative nature of the workload refers to its sheer amount, the qualitative nature refers to its task complexity and to the degree to which it uses the skills and abilities of the employee. For example, two people may both have the same quantity of work to perform in an hour. One person's workload may involve the repetition of the same task over and over again throughout the hour. The other person's workload may cover a wide variety of tasks such as making phone calls, then meeting with people, and then working in solitude on some technical problems.

Consider how poor person–environment fit on workload may lead to affective strain when persons have (a) too little complexity on their jobs, (b) too much complexity, and (c) good fit on complexity. For persons with too little complexity on their jobs, excessive workload increases the repetition or amount of unchallenging and uninteresting work. The result should be high levels of boredom, job dissatisfaction, and perhaps depression. Persons with too little workload, compared to what they want, will also be bored both because there is not enough workload and because the work is uninteresting. Persons with good fit on workload should have somewhat higher levels of job satisfaction and somewhat lower levels of boredom and related affects. Such persons at least have quantities of work to their liking even if it is not as complex as they would like.

For persons in a job with too much complexity, too much quantitative workload should tend to be cognitively exhausting, and should produce dissatisfaction and perhaps anxiety, whereas too little quantitative workload should not have this effect. Consequently, there should be a linear relationship between quantitative workload fit and job-related affects among persons with jobs that are too complex.

When the person has good fit on job complexity, then the effects of variations in workload fit on strain should be halfway between the effects found when the person has either too much or too little complexity compared to what the person wants. The resultant relationship between workload fit and strain for persons with perfect complexity fit should be a weak one, being neither linear nor U-shaped.

Methods of Analysis

Selecting Predictors

When a predictor measure was available both in the form of E_s and $(P-E)_s$, the form explaining the most variance in strain was selected on the grounds that it best explained the nature of the job demands being studied. Prior analyses (Caplan *et al.*, 1980), and an extensive study of the measurement and theory of person–environment fit using these data (Harrison, 1976; and Chapter III), showed that the measures of fit were stronger predictors than the E components in most but not all cases.

Selecting Dependent Variables

A strain was never used as its own conditioning variable because we would have been unable to distinguish between the effects due to conditioning and the effects due to restriction of range for the strain—the two effects being virtually identical. Whenever the general affects were used as conditioners, only the job-related affects were used as dependent variables. This strategy was adopted to reduce the conceptual overlap and the correlation between the conditioning and dependent variables.

As a precaution against chance findings, the hypotheses were tested on the occupationally stratified random sample, and if they were significant, an attempt was made to replicate them on a second stratified random sample drawn from the total sample.

Testing for Interaction Effects

Each predictor, except for the measures of fit, and each conditioning variable was divided into tertiles represented as the categories 'high', 'medium' and 'low'. When person–environment fit was a predictor, it was divided into the following three categories: $P > E$, $P = E$, and $P < E$. $P = E$ was defined as any score within one scale point of $P-E = 0$.

The search for interaction effects concentrated on the percentage of variance accounted for by the interaction term independent of the main effects, the main effects having already been examined and described previously (Caplan *et al.*, 1980). This search was accomplished using multiple regression analysis.

The interaction between any pair of variables was represented by a set of dummy variables (Suits, 1957) in order to include them in the *linear* multiple regression analyses. Any particular variable can be broken into dummy variables by creating a new variable for each possible value of the original variable. Each of these new variables is coded as 1 if the respondent checked off the value represented by the variable or as 0 if the person did not check off the value represented by the variable. For example, consider a measure of person–environment fit that has three scores: $P < E$, $P = E$ and $P > E$. This variable has two degrees of freedom because only the respondent's standings on any two of the scores need to be known to determine the respondents score on the variable (for example, not $P < E$ and not $P = E$, therefore, $P > E$). A dummy variable is created for each degree of freedom. For a P-E fit measure, two dummy variables (any two) would be created. Creating the third dummy variable would not only be redundant but would make the computation of any multiple regression indeterminant. As an example, suppose a person had a score of $P = E$. We had two dummy variables, $P < E$, and $P = E$. In which case, the person would receive a score of 0 on the first dummy variable and a score of 1 on the second. The sum of the variance of these two dummy variables in a multiple regression provides the total amount of variance accounted for by the original variable.

In order to represent an interaction effect as a dummy variable, one can create an analysis of variance matrix such as in Table V.1. It has two three-level predictors. The matrix describes the procedure we have followed. Each predictor, *A* and *B*, has been trichotomized. Each predictor has two degrees of freedom. The interaction effect has four degrees of freedom [(number of rows – 1) $\times$ (number of columns –1)]. Consequently we can represent the interaction effect by combinations of values representing a person's standing on any two of the rows and columns. Using Table V.1, we shall code information about the first two columns and rows to represent the four degrees of freedom. The dummy variable will consist of a

Table V.1 Matrix showing representation of two three-level variables for testing interactions

Predictor B	Predictor A		
	$P > E$	$P = E$	$P < E$
$P > E$	1	2	3
$P = E$	4	5	6
$P < E$	7	8	9

set of four-digit codes of the form *wxyz* where *w* represents whether or not the person scored P > E on the predictor *A* in Figure V.1, *x* represents whether or not the person scored P = E on predictor *A*, *y* represents the score of P > E on predictor B, and *z* represents the score of P = E on B. A code of 1 is given whenever the person *does* have a particular score on the fit variable; a code of 0 is given whenever this is *not* the case. For example, a person who scored P < E on predictors *A and B* gets a 0 for all four digits, 0000. Such a person did *not* score P > E or P = E for either predictor *A* or predictor *B*. A person who scores P > E on predictor *A* and P = *B* on predictor *B* gets a score of 1001. A person who scores P > E on both predictors *A* and *B* gets a score of 1010, and so on. There are nine such combinations representing all nine cells of the matrix in Table V.1.

In computing the actual contribution of the interaction effects in the multiple regression analysis, the interaction is entered as a set of dummy variables. The respective main effects need not be entered as sets of dummy variables. Instead they can be entered as the original trichotomized variables as was done in this case (the results would be identical whether using the original coding or using dummy variables to represent the main effects).

The computer program used for this analysis (REGRESSION in MIDAS, Fox and Guire, 1976) prints out the *F* test for the effect due to the set of dummy variables representing the interaction. This test measures the effect of the interaction holding the main effects statistically constant. The percentage of variance in strain accounted for by the interaction effect is reported as (a) the sum of squared deviations from the mean of the dependent variables due to the interaction (SSI) divided by (b) the sum of squared deviations from the mean of the dependent variable due to within-treatment variance (SSW). The sum of squared deviations of each main effect (SSB) was used to give a comparable estimate of the percentage of variance in strain accounted for by the main effects.

Whenever a significant interaction effect was uncovered by multiple regression analysis, then the means for the 3 level × 3 level analysis of variance matrix were generated. Those means were inspected to see if they conformed to the pattern of relationships that would be predicted from our hypotheses.[7] Findings that were significant but were not in the correct direction were examined with care to determine if they were due to sampling error. If the unexpected pattern of relationships did not suggest a plausible and alternative hypothesis, and if the number of such findings was about what would be expected by chance, then such a finding was not considered worth presenting in any detail.

An exception was made to the above procedure with regard to the testing of one hypothesis involving a triple interaction—the interaction of Type A, Participation, and job demands. Trichotomizing all three predictors would

---

7. The intimate relationship between multiple regression and *n*-way analysis of variance is discussed in detail by Cohen (1968), and Darlington (1968).

have produced a 27-cell analysis of variance matrix. As an alternative, the following procedure was adopted.

1. Type A and Participation were both trichotomized and then crossed to form a matrix with three levels of Type A (low, medium, and high) and three levels of Participation.
2. The correlation between job demands and the strains was run within each of the nine cells formed by the above matrix. According to the hypothesis for the triple interaction, we expected to find the strongest positive correlation between job demands and strain in the cell representing high Type As who were also low on Participation. We expected to find weaker correlations for high Type As who were medium or high on Participation and for persons in the middle tertile on Type A. We expected to find no such effect for persons labelled as Type B (low for Type A).
3. When correlations appeared to fit the pattern just described, an overall significance test of the interaction effect was conducted. This test was done using multiple regression. The predictors in the multiple regression were the main effects of Type A, Participation and the relevant job demand or stress, the two-way interactions between all pairs of these main effect variables, and the three-way interaction. The three-way interaction was represented as a multiplicative term in which the job demand or stress was multiplied by Type A which was then multiplied by the reciprocal for Participation. The reciprocal was chosen because our hypothesis is that *low* Participation increases the effect of stress on strain among high Type As.

The multiple regression analysis produces the partial correlation of each of these prediction terms independent of the other terms and provides a test of the statistical significance of each term.

Results

With regard to the four conditioning hypotheses, only one, H4, was clearly supported, and there was weak support for a second of the hypotheses. The data for the strongly supported hypothesis are described first, and then the findings regarding the other hypotheses are discussed briefly.

H4: The Qualitative Content of the Job Conditions
the Relationship between Quantitative Workload and Strain

Quantitative Workload Fit and Job Complexity Fit were used as the respective components of the interaction of predictor and conditioning variables. The measures of fit were chosen over their E components because fit was usually the stronger predictor of strain in each case.

Job Complexity Fit was believed to be the best indicator of the qualitative content of work for the following reason. Although there were other measures of job demands and fit which did refer to various qualitative dimensions of work (responsibility, ambiguity of the task, and participation for example), each of these measures already combined both the content of the demand with its quantity or load by asking the respondent to report '*How much*' there is of this quality.

The index of Job Complexity Fit, by contrast, described a series of jobs in terms of various qualitative components (contact with people, predictability of the tasks, contacts across organizational boundaries, level of variation in procedures and tasks, variance in the pace of the workload, and the number of tasks in process at any one time). The index did not emphasize the amount of quantitative load associated with the components. The respondent indicated the extent to which the job corresponded to these characteristics.

The index of Job Complexity Fit represents a conglomerate of aspects of work roles rather than some unitary concept. Nevertheless, theoretical and empirical research (Hackman and Lawler, 1971; Kohn, 1969; Quinn and Shepard, 1974) suggests that its aspects tend to be associated with one another in the design and implementation of work roles. Our own data supported this conclusion because the cross-sectional estimate of reliability for the index was .71 and the average intercorrelation among the items was .30.

Figure V.1 shows the interaction between Job Complexity Fit and Workload Fit in summary form. The interaction was statistically significant for each of the job-related measures of strain, the general affects, and Somatic Complaints with an exception to be noted below. An examination of the intercorrelations among these strains (presented in Appendix G of Caplan *et al.*, 1980) shows that the coefficients were all positive, averaged .31, and had only six values that were under .20. Consequently, to present the strongest illustration of the interaction effect, we were able to construct an overall index of psychological strain without diluting the pattern of findings.

The respective main effects of Workload Fit and Complexity Fit accounted for 10 per cent and 9 per cent of the variance in the dependent variable in Figure V.1. The interaction effect accounted for an additional 6.5 per cent of the variance (all F tests $p < .001$).

The relationship between Workload Fit and overall psychological strain was nonsignificant for persons with good fit on Job Complexity (eta $= .16$, $p = .20$) as predicted. For persons with more complexity than they wanted, strain increased steadily as Workload Fit scores changed from P > E to P = E and then to P < E (eta $= .60$, $p < .001$). For persons reporting too little complexity, the relationship between Workload Fit and strain had an increasing slope which was relatively weak and only approached significance (eta $= .23$, $p = .12$). It did not show the U-shaped curve that was predicted.

Given the possibility that the interaction effect was due to chance, we replicated its results on the second occupationally stratified, random subsample. The results of this replication are shown in Figure V.2.

Again, the interaction was statistically significant. Again, the relationship was weakest for persons with good fit on complexity (eta $= .17$, $p = .18$).

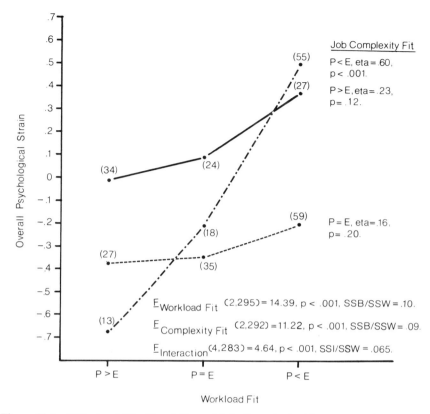

Figure V.1   Effects of Workload Fit and Job Complexity Fit on overall psychological strain (the job-related affects, general affects, and Somatic Complaints) in the first random sample

Persons with jobs more complex than they wanted showed the linear relationship between Workload Fit and psychological strain depicted in the previous figure (eta = .46, p < .001). Persons with jobs less complex than they wanted showed the predicted curvilinear relationship between Workload Fit and psychological strain (eta = .42, p < .001), a finding that did not occur in the data shown in Figure V.1.

The results shown in these two graphs suggest that misfit with regard to workload is mainly a source of strain for persons with jobs that are too complex rather than not complex enough. The predicted effect of an overly simple job on the relationship between Workload Fit and psychological strain received only partial support.

Although we had combined all the measures of psychological strain together because their effects were generally similar, we thought that the strain of Boredom might behave rather differently. Boredom should be more responsive to changes in workload in jobs with insufficient rather than

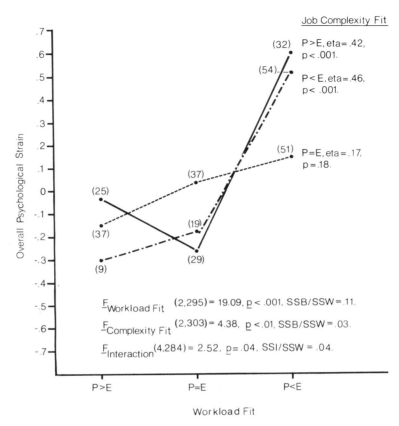

Figure V.2   Effects of Workload Fit and Job Complexity Fit on overall psychological strain. Replication on second random sample

oversufficient complexity. This prediction seemed reasonable on empirical grounds as well as theoretical grounds because Boredom correlated .51 with Complexity-Poor Fit. Boredom was greatest for persons with too little complexity. Analyses of the data, however, showed that there was no interaction effect of Job Complexity Fit and Quantitative Workload Fit on Boredom. There was only a weak main effect of Workload Fit on Boredom $(r = .13, p < .05)$. Consequently, poor fit on Job Complexity would appear to be sufficient enough to determine Boredom but not to determine the other psychological strains.

The overall impression that we draw from examining Figures V.1 and V.2 is that the conditioning effects of Complexity Fit on the relationship between Workload Fit and Strains, except for Boredom, occurred most strikingly when the person was overloaded $(P < E)$ rather than underloaded. Consequently, when there were significant changes in mean levels of strain, they tended to occur between the points of workload $P = E$ and $P < E$ rather than between $P = E$ and $P > E$.

Do Occupations with Different Levels of Job Complexity
Show Different Relationships between Workload Fit and Affective Strains?

We have just seen that Complexity Fit conditioned the relationship between Workload Fit and Affective Strains. It should be possible to show that occupations characterized by high complexity differ from those characterized by low complexity in how Workload Fit determines affective strains. We presented some evidence of this in Caplan *et al.* (1980, Table III.8) where it was noted that occupations with positive correlations between workload misfit and affective strain appeared to have poor fit on Job Complexity. If these relationships do vary by occupation, it would be a warning to avoid blindly generalizing first-order relationships among stresses and strains to all occupations (Caplan *et al.*, 1980, p.103).

In the analyses that follow, we have selected occupations on the basis of their goodness of fit on Job Complexity in order to see if they show the patterns of relationships plotted in the preceding two figures. The occupations were selected by using the following procedure. Frequency distributions for Complexity Fit were generated for each of the 23 occupations. Occupations which had at least 50 per cent of their occupants in any one fit category ($P > E$, $P = E$, $P < E$) were selected to represent that category.

Table V.2   Distributions of scores on job complexity fit given in percentages for selected occupations

Occupation	*n*	Complexity P > E	P = E	P < E
Machine-paced assemblers	77	48.1	27.3	24.7
Police	110	10.9	60.0	29.1
White-collar supervisors	42	9.5	60.9	28.6
Train dispatchers	82	17.1	30.5	52.4

Table V.2 presents the occupations chosen as most representative of each of the types of P–E fit categories for job complexity. The percentage of respondents by category of fit is given for each occupation. Machine-paced assemblers were selected to represent jobs with too little complexity ($P > E$). Police and white-collar supervisors were selected to represent occupations tending to have a good fit ($P = E$). Train dispatchers were selected to represent an occupation with too much complexity ($P < E$). All of the other occupations in the study either showed no substantial majority of employees in any one PE fit category or had sample sizes which were too small (less than 30 employees) for the analyses to be considered reliable.

Figure V.3 presents plots of the relationship between Workload Fit and the measure of overall psychological strain for each of the four selected occupations. The pattern of findings is very similar to that presented in Figures V.1 and V.2. The occupations with good fit on complexity (police and white-collar supervisors) showed weak relationships of Workload Fit to strain (etas = .27

84

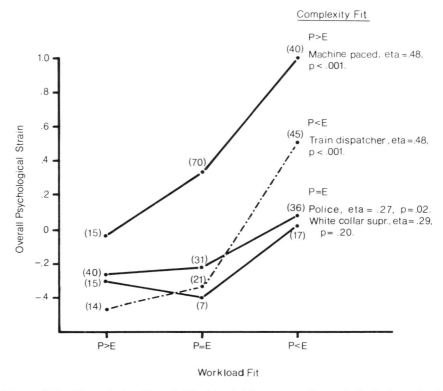

Figure V.3 The relationship of Workload Fit to overall psychological strain for selected occupations. Etas follow each occupation's name. *N*'s in parentheses

and .29 respectively). Both of the occupations with misfit on complexity, however, showed strong associations between Workload Fit and strain (etas of .48). The major effect of Complexity Fit on strain was again present for persons with too much rather than too little workload.

Implications of these findings for preventive occupational health. The preceding findings are of particular interest because they suggest some steps which might be taken in programs of preventive occupational health. One may want to identify employees who have misfit on Job Complexity and high Workload. These employees may be most likely to develop the highest levels of psychological strain compared to employees with good fit on Job Complexity or with good fit or underload on Quantitative Workload. It then becomes a matter of deciding whether or not the employee at relatively high risk of psychological strain can be helped best by either changing fit on workload or fit on job complexity. The factors that would influence such a decision include the practical ability of the organization to alter the workload, the job complexity, or the person. These choices are discussed further in the last chapter.

Before an organization could consider such alternatives, it would need to conduct preliminary measurements of person–environment fit. Certain occupational groups may be more likely than others to experience both complexity misfit and quantitative work overload. Preventive and ameliorative efforts can be concentrated among the occupations at greatest risk. The analyses presented in Figure V.3 show that occupational differences do occur. Each organization would need to determine whether or not such occupational differences were present in its settings.

Nonsignificant and Marginal Findings

There were only a few significant findings which represented conditioning effects of psychological strain, personality, or participation on the relationship between stress and strain. These significant findings were about as frequent as one would expect by chance. Furthermore, the majority of the significant interactions did not conform to our hypotheses and were not readily interpretable.

One of the hypotheses did produce one pattern of results that was in the expected direction, but the support is weak, and replication is required. We are referring to the hypothesis that the ability of job stress to produce strain would be greatest for Type A persons who lacked participation. Of 128 tests of this prediction using psychological strains as the dependent variables, one test produced a significant triple interaction — a result at chance level. The finding is presented in Table V.3.

The analysis was conducted separately for white-collar employees and for blue-collar employees. A very large sample was needed for this analysis because it tests the interaction of four variables (Type A, Participation, Job Complexity-Poor Fit, and white- versus blue-collar). As a result all blue-collar employees were pooled (the first through ninth occupations listed in Table II.1, page 10). Similarly all white-collar employees were pooled (the tenth through 23rd occupations). Furthermore, in order to have sample is of as equal size as possible within each tertile of Type A and within each tertile of Participation, the tertile split points were determined separately for blue and for white-collar employees within their respective distributions. The implications of this for interpreting the results is discussed below after the results are described.

For white-collar employees the highest association between Job Complexity Poor Fit and Anxiety was for persons with the highest Type A scores and the lowest Participation ($r = .37, p < .001$). The correlations decreased for high Type As with medium and high levels of Participation. The correlations between stress and strain were lower for the other levels of Type A and showed no such decrease. The interaction is significant at $p = .01$. For blue-collar employees, those who were Type A and had the lowest levels of Participation again showed the highest correlation between stress and strain ($r = .45, p < .01$), but here the replication of white-collar results ends. Medium Type As with high Participation had a stress–strain relationship that was almost as high ($r = .41, p < .01$).

Table V.3 Effects of Job Complexity Poor Fit on anxiety as a function of Type A and Participation

| | Vickers Type A | | | | | | | | |
| | Low | | | Medium | | | High | | |
Participation	Low	Medium	High	Low	Medium	High	Low	Medium	High
White collar									
r	.04	.27[b]	.26[b]	.22[a]	.29[b]	.15[a]	.37[c]	.31[b]	.15[a]
n	123	65	128	116	166	174	103	167	174
Blue collar									
r	.22	.11	.10	.15	.15	.41[b]	.45[b]	.10	.25[b]
n	60	30	66	56	37	87	42	44	96

Note

| | White collar | | | Blue collar | | |
Effect	Partial r	t	p =	Partial r	t	p =
1. Complexity–Poor Fit	−.06	−2.13	.03	−.03	−.58	n.s.
2. Type A	−.06	−2.30	.02	−.04	−.87	n.s.
3. Low Participation	−.07	−2.68	.008	−.01	−.25	n.s.
4. 1 × 2	.07	2.45	.01	.05	1.08	n.s.
5. 1 × 3	.07	2.76	.006	.01	.26	n.s.
6. 2 × 3	.07	2.51	.01	.01	.24	n.s.
7. 1 × 2 × 3	−.07	2.60	.01	−.01	−.31	n.s.

$r = .24, p < .001, n = 1374$ $r = .31, p < .001, n = 518$

[a] $p < .05$.
[b] $p < .01$.
[c] $p < .001$.

Although there were some predicted patterns of interaction when we examined the effects on the physiological dependent variables, none of these patterns turned out to be statistically significant.

The lack of significant findings among the blue-collar workers may be attributable in part to the fact that the points that were used to split them into tertiles on Type A and on Participation were different from those used for the white-collar employees. Had we used the same cut points for both occupations, there would have been few blue-collar employees with high Participation and with high scores on the Vickers Type A measure.

We can suggest that the presence of significant interaction effects among the white-collar workers and not among the blue-collar workers lies in the differences in their distributions on the predictor variables. White-collar employees have the greatest range of scores on Participation, Complexity Fit, and Type A. The nature of the persons and their jobs makes this so. It is a substantive difference and not merely an artifact of the data. It suggests that the study of Type A traits is more relevant in white collar than in blue-collar occupations.

The failure of our analyses to confirm many of the interaction hypotheses may have occurred either because the data genuinely disconfirmed the theoretical hypotheses, or because the hypotheses did not receive adequate tests due to invalidity of the measures and the inappropriateness of the research design. For example, studying acute rather than chronic stressors and strain would reduce the chance of stabilized correlations among stressors, their moderators, and the dependent strains and would increase the chances of uncovering more conditioning effects. In the next section we discuss the heterogeneous nature of our samples as one possible cause of the non-significant findings.

Conditioning effects of occupation on interaction effects—another triple interaction. Some exploratory analyses were conducted to see if the hypotheses described at the beginning of this chapter would be confirmed in replication from one occupation to another. In order to obtain adequate sample sizes for occupational groups, and in order to study a small number of occupational groups that were relatively different from one another in their job character-istics, the following samples were chosen. We selected administrators, a group which had a large sample size and represents professional white-collar workers. We combined air traffic controllers from small and large airports as a second group because there were few differences due to size of airport; combined, they produced an adequate sample for analyses. We felt that the responsibilities of the controllers and the nature of their work made them a unique group for study. Finally, we formed a blue-collar sample which was made up of machine-paced and nonpaced assembly line workers, relief workers on assembly lines, toolmakers and diemakers, continuous-flow monitors and forklift drivers. Admittedly, the blue-collar group represents a diverse set of occupations. The diversity was necessitated by our desire to have a blue-collar group and by our need to have a large sample size in order to examine interaction effects within the sample.

The results of our search for interactions within these occupational groups suggested that whether or not a variable such as personality or affective state conditions the relationship between stress and strain may depend on the occupation that is examined.

For example, for air traffic controllers, Responsibility for Persons was associated with high levels of Depression only if level of Participation in decision-making was low. If Participation was medium, there was no relation-ship between Responsibility and Depression. If Participation was high, then responsibility appeared to be a source of relief from depression so that the higher the Responsibility the lower the level of Depression ($F_{\text{interaction}} = 3.33$, $df = (4,115)$, $p = .01$). This effect was not observed among administrators or blue-collar workers.

To take another example, misfit on Job Complexity was more strongly related to high Job Dissatisfaction for highly anxious blue-collar workers, and was less strongly related for each group of successively less anxious blue-collar workers ($F_{interaction}$ = 2.73, df = (4228), p = .03). No such finding was evident for the air traffic controllers or the administrators.

As a final example, misfit on Job Complexity was strongly associated with Job Dissatisfaction for administrators who reported low Participation in decision-making (a U-shaped fit curve was produced with lowest Dissatisfaction occurring when P = E). For administrators who reported medium Partici-pation, the effect was less marked and was even weaker for those who reported high Participation ($F_{interaction}$ = 2.78, df = (4,239), p = .03). No such effects were found for the other occupations examined.

It is quite possible that these findings are due to chance, for we have presented only those significant findings which made some sense. The total number of significant findings at $p < .05$ is about what you would expect by chance. So, further replication is needed.

To gain some insight into why the findings held for some occupations and not for others, we examined the means and standard deviations of the predictor, conditioning, and dependent variables. Perhaps the absence of findings occurred in occupations where there was a restriction of range on these variables, as indicated by a low standard deviation. Perhaps the positive findings only occurred in occupations above a certain mean level on these variables.

Table V.4 shows the means and standard deviations for the blue-collar workers, air traffic controllers, and administrators for the variables that have just been considered. There were significant occupational differences in the means and variances for almost every variable. The only exceptions were that there were no differences for Anxiety, and the difference in means between air traffic controllers, and administrators on Job Dissatisfaction was nonsignifi-cant. So it is quite possible that the occupational differences in mean levels and/or the variances of the predictors, conditioning, and dependent variables are a partial explanation for why some of the interaction effects appeared in one occupation group and not in the others. However, the significant effects were no more likely to occur for occupations with the smaller variance on the variables than for occupations with larger variances. Furthermore, an inspection of the means and standard deviations does not suggest any theoretical explanation for the occupational differences in the interaction effects.

Aside from preceding observations, let us assume, for the sake of speculation, that these occupational differences in effects of job demands on strain were not due to chance. Then the results would suggest that it would be difficult to detect similar interaction effects in the stratified random sample. The occupational differences in the degree to which interactions were present would either cancel out or water down any overall evidence of interactions.

Even if these interactions do exist for one occupational group and not for another, their use in preventive programs of occupational health would be uncertain. As we show in Chapter VI, occupation in this study is really a surrogate for a variety of characteristics of the job and of the person, many of

Table V.4   Differences in means and variances for selected occupational groupings on selected variables

Variable	Blue collar ($N = 354$)		ATC ($N = 124$)		Administrator ($N = 253$)		F^a_{means}	$F^b_{variances}$
	X	SD	X	SD	X	SD		
Responsibility–E	2.39	1.17	2.94	1.14	2.53	1.00	80.07	4.02
Participation	2.49	1.13	3.14	.99	3.39	.85	60.43	10.85
Job Complexity Poor Fit	.18	6.73	1.78	4.30	3.03	3.79	19.86	49.30
Job Dissatisfaction	3.83	.82	3.00	.68	3.10	.75	89.02	3.52
Anxiety	1.66	.57	1.72	.55	1.72	.50	1.22	2.80
Depression	1.72	.56	1.50	.41	1.61	.50	9.28	8.01

Note   The blue-collar category includes machine-paced and nonpaced assemblers, relief workers on assembly lines, toolmakers and diemakers, continuous-flow monitors, and forklift drivers. ATC = Air traffic controller.

[a] F test of occupational differences in means. Only the F test for Anxiety was nonsignificant.

[b] F test of occupational differences in variances according to Box (1949). Only the F test for anxiety was nonsignificant.

which are represented by variables measured by our study. Consequently, if an effect holds only for air traffic controllers, we need to know which underlying characteristics of air traffic control as an occupation and of the controllers as role occupants, out of the total pool of variables which distinguish air traffic control from other occupations, are the relevant determinants of the interaction. Laying out hypotheses regarding triple and higher-order interactions would be a necessary next step before analyses could be pursued with confidence.

Nonsignificant findings as support of person–environment fit theory. The lack of significant findings regarding the interaction of job demands and either personality traits or emotional states of the person may provide partial support for a major hypothesis in person–environment fit theory. P–E fit theory states that the interaction effects are most likely to be significant when characteristics of the person and of the environment are measured on commensurate dimensions (for example, amount of role ambiguity on the job and tolerance for such ambiguity—Frenkel-Brunswik, 1949). In this chapter, the use of noncommensurate dimensions of person and job environment characteristics has proven largely unsuccessful as a predictor of strain. By contrast, the use of commensurate dimensions in Chapter III has been much more successful. The internal reliabilities of the measures of the person used in Chapter V were about as high as the internal reliabilities of the P measures used for the P–E fit dimensions (and the same measures of E and strain were used) so it is unlikely that the lack of results with noncommensurate measures was due to low index reliability. The conditioning variables chosen in this chapter to represent measures of the person could have been irrelevant theoretically. That irrelevance, of course, is the very thing that P–E fit theory hopes to eliminate by the use of commensurate dimensions of P and E.

Chapter VI

Person and Environment versus Occupation as Predictors of Employee Strain

In Chapters III through V we described a number of relationships between job demands or stress and employee wellbeing. In this chapter we examine the extent to which occupational groups differ in ways that mirror these stress–strain relationships. Then we consider the merits of using occupation versus our measures of stress as predictors of employee wellbeing.

We will suggest that merely knowing that some occupations are high in mental and physical strain and others are low is not enough to improve the nature of work environments. For one thing, symptoms often fail to reveal their causes. Second, occupation categories may represent a multitude of job demands of which only some demands may determine employee wellbeing. One will need to know these determinants to launch an effective program of preventive health in work organizations.

Occupational Differences in Stress and Strain

Strain

Which occupations had the highest levels of strain? A very detailed presentation of the data on this question appears in Caplan *et al.* (1980). Table VI.1 presents a summary of occupational differences in strain. For the purposes of simplifying the presentation, the 23 occupations have been collapsed into four groups. The unskilled blue-collar jobs include assembly line workers, relief workers on the lines, forklift drivers, and machine tenders. The skilled blue-collar workers are represented by toolmakers and diemakers. The white-collar nonprofessionals include police, electronic technicians, foremen and supervisors, air traffic controllers, and train dispatchers as well as others. Finally, the professionals include scientists, physicians, professors, administrative professors, administrators, accountants, and engineers. Appendix A (page 125) presents a summary of occupational differences with data for each of the 23 occupations.

As can be seen from Table VI.1, blue-collar workers were high on Job and

Table VI.1 Summary of occupational differences in strain

| Strain | Blue collar | | White collar | |
	Unskilled	Skilled	Nonprofessional	Professional
Workload Dissatisfaction	+	0	0	0
Job Dissatisfaction	+	+	–	–
Boredom	+ +	0	–	–
Anxiety	0	0	0	0
Depression	+	0	0	0
Somatic Complaints	+ +	0	0	0

Note Each + or – represents one-third of a standard deviation from the mean of the stratified random sample; 0 = near the mean.

Workload Dissatisfaction and white-collar workers were low. Unskilled blue-collar workers, compared to other groups, had very high scores on Boredom, on Depression, and on Somatic Complaints. Among these unskilled blue-collar workers, workers on machine-paced assembly lines had very high levels of strain. White-collar employees had particularly low scores on Boredom. There were no occupational differences in levels of anxiety.

For the subsample for which we have physiological data, we found that the scientists, compared to all six other occupational groups, had the lowest level of systolic blood pressure. This was the case even when the data were stratified by age groups. This finding is presented in Figure VI.1. There was a similar but nonsignificant trend for diastolic blood pressure. These findings parallel those

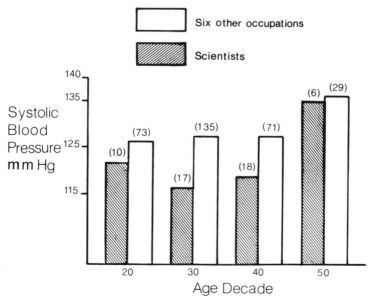

Figure VI.1 Occupational differences in Systolic Blood Pressure $F_{(6352)} = 3.78$, $p < .002$ across occupations

found at NASA where scientists, compared with administrators and engineers, had the lowest levels of blood pressure and of stress (Caplan, 1972). There were no occupational differences in cholesterol, uric acid, heart rate, or other physiological variables.

As part of the study we asked people the reasons for their last visit to a physician for other than a routine check-up. These data produced the following results. An analysis by occupation showed that cardiovascular disease (chi^2 = 36.83, d.f. = 22, p = .03) and respiratory diseases such as colds (chi^2 = 54.77, d.f. = 22, p < .001) differed significantly in their reported frequencies across the occupations. Professors had the lowest rate for both of these diseases. We were unable to draw further conclusions about occupational differences because the sample sizes were too small for this purpose.

Were the High Strain Occupations also High in Stress?

If stress causes strain, the answer to this question should be 'yes'. Table VI.2 summarizes occupational differences in levels of job stress. When we examined occupational differences in stress, the unskilled blue-collar workers stood out. They had the greatest underutilization of their skills and abilities and the poorest person–environment fit between desired and the experienced amounts of complexity, responsibility and role ambiguity.

Table VI.2 Summary of occupational differences in stress

Stress	Blue collar		White collar	
	Unskilled	Skilled	Nonprofessional	Professional
Underutilization	+ +	–	0	—
Poor Fit				
Complexity	+ +	0	0	—
Responsibility	+	+	0	—
Role Ambiguity	+	0	0	0
Workload-Excess	0	0	–	+
Future Ambiguity	+ +	+	0	0
Low Participation	+	0	0	0
Low Social Support	+	0	0	0

Note Each + or – represents one-third of a standard deviation from the mean of the stratified random sample; 0 = near the mean. The modal score across the occupations in each category has been used for the summary.

Although unskilled blue-collar workers had poor fit on the qualitative aspects of their work, they were about average on Quantitative Workload Fit. It was the professionals, particularly administrative professors, who had the highest levels of Quantitative Workload. Judging from the occupational differences in psychological strain, Quantitative Workload was a less important stress across the occupations than misfit on the qualitative nature of that workload. As reported in Chapter IV, overall Job Dissatisfaction was more likely to be determined by these qualitative aspects of the job than by the workload *per se.* Moreover, Chapter V showed that the effect of Quantitative Workload Fit on strain was conditioned by the complexity of the work; misfit with regard to workload was a source of strain mainly for employees whose jobs were too complex rather than too simple.

Unskilled blue-collar workers also had very high levels of ambiguity about the future security of their jobs and the future value of their job skills. Furthermore, they reported low Participation and low Social Support from Others on the job. In particular, workers on machine-paced assembly lines reported high levels of job stress.

By contrast, the skilled blue-collar and professional groups reported little Underutilization. The professionals reported better than average fit between the amounts of Complexity and Responsibility for Persons on the job and the amounts they wanted. Nonprofessional white-collar workers fell near the mean on all the stresses in Table VI.2.

These occupational level findings, like the individual level findings in Chapters III through V, suggest that stress and strain are related because the blue-collar unskilled workers had high levels of both strain and job stress, whereas stress and strain tended to be low for white-collar professionals. There are, however, a number of other possible interpretations of these associations between occupation and either job demands or psychological strains. Some occupations may tend to attract employees who perceive their jobs as demanding or who tend to be psychologically strained prior to entering the job. Employees who are more sensitive to job stress may be attracted to some jobs and not others. Such attraction can occur through employee self-selection into the job and employer selection of employees with such traits. Osipow (1968) details a number of these theories of occupational choice.

Some employees may be selected out of some jobs by promotion, transfer, and quitting. If selecting out is more likely to occur in some jobs than in others, then the most misfit and the most dissatisfaction with a job will tend to occur in occupations where selecting out does not occur. Occupations requiring a good deal of training, such as the job of family physician, may select out many of the ill-fitting aspirants to the occupation during the training process. Occupations which require little training, such as work on an assembly line, may have a wider range of performances which are tolerated before selection out occurs. Consequently, occupations vary in the amount of misfit that can be tolerated.

Socialization into the occupation is another way in which occupational differences in misfit and psychological strain can be influenced. Occupations with little turnover, either because alternative job markets are poor, or because the incentives for remaining in the job are good, may provide more opportunity for the employee to become socialized to the demands of the job and perhaps even to the acceptable level of satisfaction that one should feel on the job. Our research design is not sufficient to test these various hypotheses because the data are cross-sectional. Nevertheless, it is important for the reader to keep these alternative interpretations in mind when evaluating the results.

Occupation as an Explained versus an Explanatory Variable: Hypotheses and Analyses

The analyses in this section attempt to determine how much of the association between occupation and levels of strain can be explained by the job demands and characteristics of the employees associated with these occupations. If measures of employee and job characteristics do appear to be important variables intervening between occupational category and strain, then occupation would represent an abstract surrogate for such mediating variables. If occupational category is unable to explain variance in strain beyond that accounted for by job demands and employee characteristics, we will have evidence that we have identified a rather exhaustive set of descriptors of these occupations and their occupants.

Each of the classes of potentially mediating variables is reviewed below. Next we present the methods used to test the preceding arguments. Following this is a discussion of the findings and their implications for programs of preventive industrial and organizational health. The actual variables used in the analyses are presented in Table VI.3, footnote 6.

Hypotheses

Characteristics of the job environment mediate the effects of occupation. Our earlier analyses (Caplan *et al.*, 1980) demonstrated significant occupational differences across the 23 occupations on every job demand examined. Other studies, including a survey of the work force of the United States (Quinn and Shepard, 1974), have also demonstrated such occupational differences. Accordingly, job demands do distinguish occupations from one another. This is hardly a profound conclusion. Nevertheless it suggests that the subjective reports of job demands reflect objective job conditions. The differences in job demands among the occupations correspond to our impressions of the objective demands of these jobs, and so we are comfortable with the use of self-reports as indicators of objective job environments in the analyses that follow.

Characteristics of the employee including traits, needs, and abilities mediate the effects of occupation. The self- and organizational selection of employees into occupations is an area of study by itself. Our earlier analyses (Caplan *et al.*, 1980) we reported significant differences in personality across the 23 occupations in the sample. For example, administrative professors (professors with administrative roles) had much higher scores on Type A personality than did professors; assembly line workers had very low scores on Type A. Consequently, predisposing factors of the employees must be considered as possible predictors of employee health.

Goodness of fit between the person's abilities and traits and the environment's demands and supplies (or rewards) mediate the effects of occupation. Goodness of fit is expected to vary by occupation because selection criteria and management practices may be more effective in some occupations than in others. In Table VI.2 (and in Caplan *et al.*, 1980) we reported clear differences in the degree of person–environment fit across the 23 occupations. Unskilled blue-collar workers, for example, tended to have poor fit on job complexity whereas white-collar professionals tended to have good fit. Both skilled and unskilled blue-collar employees had poor fit on responsibility for persons but white-collar professionals had good fit. The amount of misfit experienced by white-collar nonprofessionals on these dimensions tended to be in between the poor and good fit of the preceding groups.

Methods of Analysis

We considered only the major correlates of the psychological strains and Somatic Complaints — those with coefficients equal to or greater than .30. These correlates were subjective measures of job demands, of the person's needs and traits, and of person–environment fit. The total set of these predictors formed the measures that have been used to determine how much variance in the strains was explained by these mediating variables and how much additional variance was explained by occupation.

As a first step, a procedure called dummy variable transformation (Cohen, 1968) was used to examine the first-order effects of occupation in a regression analysis. Occupation was coded into 23 categories, one for each occupation. Then it was transformed into 22 dummy variables for 23 minus 1 degrees of freedom in the original variable. Each employee was assigned a score of 1 or 0 on each of the dummy variables, 1 representing 'a member of this occupational category' and 0 representing 'not a member of this occupational category'. Thus 22 dichotomous scale variables were created for the analyses. R^2, for these 22 predictor variables as a set, represents the percentage of variance accounted for by occupation. There is a detailed description of similar dummy variable coding of conditioning variables in Chapter V, p.77.

For the second step we determined the percentage of the variance in strain that was accounted for by the set of stresses and strains hypothesized to intervene between occupation and the strain. To do this, the hypothesized intervening predictors were entered into a multiple regression analysis. R^2 provided the total percentage of variance accounted for by the hypothesized mediating predictors. Multiple regression analyses were performed for each index of psychological strain and for Somatic Complaints as dependent variables. Although we have shown that the types of stresses and measures of the person which are related to strain depend on the type of strain, the same total set of predictors was used for computing the multiple regression on each strain. The use of a standard set of predictors allows the reader to compare the relative contribution of the same predictors with regard to each strain.

Dummy variable coding was also used to enter an interaction effect as a predictor in our analyses of the mediating variables when that interaction was not between an index of P and its commensurate E index. Such terms were formed only when previous analyses showed a significant interaction. The interaction between two measures of fit, Job Complexity Poor Fit and Workload Fit, was the only such significant effect on psychological strain. Chapter V describes that interaction and the procedure used to represent this interaction as a dummy variable in the regression analysis.

R^2 and R^2-adjusted-for-degrees-of-freedom were virtually the same because of the large sample size. Consequently only R^2 has been reported.

Findings

Three questions were asked of the data.
(1) What percentage of the variance in measures of psychological strain (affective state) was accounted for by occupation as the sole predictor?
(2) What percentage of the variance in these strains was accounted for by the measures of role demands or environment, needs or preferences, personality traits, and person–environment fit as predictors?
(3) What percentage of the variance in the strains was accounted for by occupation as a predictor beyond the effects of role demands, personality and fit?

Table VI.3 presents the results for these three questions.

(1) Occupation as the sole predictor. In column A of Table VI.3 it can be seen that occupation accounted for only 4 to 7 per cent of the variance in Anxiety, Depression, Irritation, and Somatic Complaints. These are strains which are not specifically related to work in their item content. On the other hand, occupation accounted for 9 to 29 per cent of the variance in the job related affects of Job Dissatisfaction, Boredom, and Workload Dissatisfaction. Analyses presented in Chapter IV would lead us to expect this pattern of findings. Those earlier findings, based on partial correlation techniques, suggested that role demands predicted to job-related affects which in turn predicted to more general affects such as anxiety and depression. Based on these results, we would expect the multiple regression of all the measures of job demands, fit, and personality should account for more variance in the job-related affects than in the more general affects, a prediction which takes us to the second of the three questions.

(2) Environment and person as mediators. Column B in Table VI.3 presents the percentage of variance accounted for by the hypothesized mediating variables. The findings show that the measures of role demands, needs, traits, and fit accounted for more of the variance in the job related affects (37 to 45 per cent of the variance) than in the general affects and Somatic Complaints (14 to 26 per cent). Second, without exception, the measures of job environment, needs, traits, and fit together accounted for substantially more of the variance in the strains than did occupation alone (mean $R^2 = .30$ versus mean $R^2 = .11$ respectively; difference significant at $p < .001$).

Table VI.3   Contributions of occupation, stress, and personality to variance in strain

			Predictors		
Strain	A. R^2 for occupation[a] only	B. R^2 for E_s, P_s, and $(PE)_s$[b]	C. R^2 for occupation plus E_s, P_s, and $(PE)_s$	D. Squared part correlation of occupation[c]	E. % of variance due to occupation explained by E_s, P_s, and $(PE)_s$
Job-related affects					
Job Dissatisfaction	.16	.38	.44	.06	62.5
Boredom	.29	.45	.51	.06	79.3
Workload Dissatisfaction	.09	.37	.40	.03	66.7
General affects					
Anxiety	.04	.19	.21	.02	50.0
Depression	.05	.26	.28	.02	60.0
Irritation	.04	.23	.26	.03	25.0
Somatic Complaints	.07	.14	.18	.04	42.8

Note $N \simeq 1600$–1860. R^2 adjusted for shrinkage is almost the same as R^2, given the large sample size, and is not presented in the table. All R^2 values in the table are significant at $p \leq .001$.

[a] Number of predictors = 22 dummy variables representing the 23 occupational categories.

[b] Number of predictors = 19: Role Conflict, Participation, Variance in Workload, Future Job Ambiguity, Underutilization, Social Support of Supervisor, Social Support of Others at Work, Fit on Complexity of Job, on Responsibility for Persons, on Role Ambiguity, on Workload, Type A Personality, Assert Good Self, Deny Bad Self, and Flexibility (the latter four are considered traits), all constitute one predictor each. The interaction of Workload Fit by Complexity-Poor Fit is represented as four dummy variables.

[c] Squared part = (R^2 with everything in) – R^2 omitting occupation), referred to by some writers as the squared semi-partial correlation and by Darlington (1968) as the 'usefulness' measure.

(3) Occupation as a unique explanatory variable. Column C indicates the percentage of variance explained in the strains when occupation and hypothesized mediating variables are used together as predictors. The difference between the percentage of variance in column C and the percentage in column B is presented in column D as the squared part correlation of occupation with the strains. This coefficient represents the ability of occupation to explain variance in the strains above and beyond that already explained by the underlying variables (Andrews *et al.*, 1973; Darlington, 1968).

It can be seen that the addition of occupation as a predictor accounted for only 2 to 6 per cent additional variance beyond the 14 to 45 per cent of the variance accounted for by the hypothesized mediating variables. In all cases, occupation added a significant amount of variance to the R^2 values, although there was not much difference from one strain to the other in the percentage of additional variance provided by occupation. Nevertheless, occupation still tended to account for more variance in the job-related affects (average $R^2 = .05$) than in the general affects (average $R^2 = .03$).

As another way of examining the data in Table VI.3, we can ask what percentage of the variance due to occupation as a first-order predictor (the first column of data) was accounted for by the job demands, needs, personality, and fit variables. To calculate this figure one subtracts 1.00 from the ratio of (a) the squared part correlation of occupation (column D) to the squared multiple correlation of occupation (column A). The results are shown in column E.

From 62.5 to 79.3 per cent of the variance in the *job-related* affects initially attributed to occupation was explained by the mediating variables. From 25 to 60 per cent of the variance in the *general* affects and Somatic Complaints initially attributed to occupation was explained by the mediating variables. These results suggest that occupation is more likely to represent unmeasured variables that predict to general affects than to job-related affects. This finding is discussed in the next section.

Discussion

The preceding analysis was a conservative test of the assertion that occupation is a surrogate for underlying measures of person needs and preferences, environment, and person–environment fit. It was conservative because no attempt was made to measure all relevant descriptors of the job environments, of the skills, abilities and personality traits of the persons, and of the nonwork environment correlates of occupation. Furthermore, the interactions among these variables were by no means considered exhaustively. Consequently, the ratios of variance due to underlying variables to variance due to occupation-plus-underlying variables are probably underestimates. They probably underestimate the extent to which underlying measures can account for occupational differences in strain. The pattern of these findings, nevertheless, does generalize beyond our particular sample. As the next section shows, our results are similar in many ways to those obtained in a nationwide occupationally stratified random sample of the workforce of the United States.

Comparison with a National Sample

In a nationwide random sample study of the work force, Quinn and his colleagues (Barnowe, Mangione and Quinn, 1973) tested a similar set of hypotheses. They began by suggesting that demographic and occupational categories had very little to offer as explanations of variance in job satisfaction after one had already considered various measures of the job. They referred to these measures as the Quality of Employment.

Table VI.4   Multiple correlations between Job Satisfaction and seven combinations of quality of employment, demographic, and occupational predictor sets ($N = 1327$ wage-and-salaried workers)

Models as defined by predictor sets	R^{2a}
Quality of employment only	.28
Occupation classification only	.05
Demographic characteristics only	.06
Demographic characteristics and occupational classifications	.08
Quality of employment and occupational classifications	.28
Quality of employment and demographic characteristics	.29
Quality of employment, demographic characteristics, and occupational classifications	.30

Note   Source is Barnowe, Mangione and Quinn (1973). R^2s reported are the values adjusted by multiple classification analysis (Andrews, *et al.*, 1973) to correct for possible capitalization on chance. Unadjusted values were somewhat higher.

Table VI.4 summarizes their findings. They used Facet-Free-Job Satisfaction, as their index of strain. That measure is very similar to Job Dissatisfaction in our study. Using multiple classification analysis (MCA; Andrews *et al.*, 1973), a method closely related to multiple regression, Quinn and colleagues found that 28 per cent of the variance in Facet-Free Job Satisfaction was accounted for by the measures of quality of employment alone. When occupation was entered into the analysis as an additional predictor,[8] no additional percentage of variance in job satisfaction was explained. And when the demographic variables of race (white or black), age, sex, and highest level of education attained were added, only two more additional percent variance were explained. In this regard, then, our findings appear to be replicated by this national sample survey.

---

8. Occupation was defined as a combination index made up of collar color, excluding farm workers, occupational status based on Duncan's prestige ratings (Reiss *et al.*, 1961), type of industry according to the nine census categories and major occupational group coded according to eleven census categories.

Nevertheless, there are some important differences in the magnitude of the findings when we compare the two studies. Our measure of occupation accounted for 16 per cent of the variance in job satisfaction whereas the Quinn *et al.* measure accounted for only 5 per cent. This is probably the case because we are using a more refined classification — the actual occupational title rather than the broader categorization using collar color, occupational status, type of industry, and related census occupation groupings. The mediating variables in our study also accounted for more variance in job satisfaction (38 per cent) than was the case in the national study (29 per cent). This probably occurred because we have more refined measures of stress at work in the sense that we have taken into account person–environment fit as well as the perceived amount of environmental demand. These findings reinforce the conclusion in Chapter III, that one improves significantly the ability to predict employee strain and wellbeing if one takes into account characteristics of the person as well as the person's work environment.

How Stress off the Job Contributes to Variance in Measures of Strain

The analysis in Table VI.3 showed that occupation did not account for variance in our measures of *general* affective states as well as it did for variance in the measures of *job-related* affects. Although these measures of strain varied by occupation, their variance may have been partly due to sources of stress off the job that could have been distributed across the occupations in a very nonrandom manner. For example, the physician may enjoy better conditions of housing, recreation, nutrition, and health than the unskilled worker because of the higher income and purchasing power of physicians. This could produce an association between occupation and general affects after the measures of job demands, personality, and fit were partialled out statistically from the relationship. Previous studies support this point. In a study of Detroit workers (Kornhauser, 1965), unskilled blue-collar workers were found to have a markedly poorer quality of life compared to persons from higher-skilled, higher-status occupations. Similarly, national sample survey data (Campbell, Converse and Rodgers, 1976) also suggest that there are important differences in the degree of nonwork stress for members of different occupational groups. One must consider whether or not such nonwork stresses have effects that spill over into the work environment and affect work attitudes and emotions. Elements of the work environment may also spill over and help determine nonwork life (for example, Kantor, 1977; Kessler, 1979).

As was noted earlier, we did not specifically examine the contribution of nonwork stress to our measures of strain in this study. There was one exception, however. We measured social support from persons at work and from nonwork sources referred to as 'wife, friends, and relatives'.

When we examined the correlations between these different sources of social support on the affects (Caplan *et al.*, 1980), the following two patterns of results appeared. First, an average of 8.1 per cent of variance in the job-related affects was explained by social support from persons at work. An average of only 4.5 per cent of the variance in the general affects was explained by social support from persons at work. Second, social support from nonwork sources accounted for an average of 1 per cent of the variance in job-related affects and 1 per cent of the variance in the general affects.

These data and other findings presented earlier provide partial support for the suggestion that job-related rather than nonwork environments tend to determine job-related affects and that job-related environments are less likely to determine more general affective states. Other studies, however, show that social support from nonwork sources, such as the spouse, do have an effect on more general affects and on physical health. Kasl and Cobb (1977), in a longitudinal study of plant closings, found that social support from the spouse was an important buffer of the effects of job loss on general affects such as anxiety and depression, on joint-swelling, and on the extent to which serum cholesterol returned to normal levels after termination of the job. Social support from spouse also had direct, main effects on depression. House (1980), in a cross-sectional study of factory workers, found that the inverse association between job stresses and life satisfaction was buffered by social support from both the spouse and the supervisor. Similar effects were found for self-reported symptoms of ulcers. Although we initially found no evidence that off-the-job and on-the-job social support buffers the effects of job demands on worker health (Pinneau, 1975), a reanalysis of these data by LaRocco, House and French (1980) uncovered support for the buffering hypothesis.

Implications for Interventions

The results clearly suggest that there is little to be gained by persisting to measure occupation as the sole predictor of strain when one's intentions are to improve employee health. For one thing, occupational categories and other demographic characteristics are not very open to change in a program of preventive occupational health. Mediating characteristics of the job and job fit on dimensions such as role ambiguity, complexity, workload, or responsibility are much more open to change. Findings about the distribution of strains across occupations may help one identify a group at risk of illness at a particular point in time, but may not apply to another point in time or at other sites. The job demands and personality traits characterizing any occupation and its occupants may also change over time. So it is more sensible to concentrate on the generalizable relationships between basic characteristics of role demands, employee attributes, and strain.

Chapter VII

Summary and Implications

This chapter summarizes the main findings of the study, points out some implications for future research, and discusses possible applications of the findings in the work place.

Summary of Findings

Objective versus Subjective Stress

Do subjective reports correspond to the objective conditions of work? Should one, for example, decide that the objective workload should be reduced when one finds a high subjective workload? Our theoretical model starts with objective stresses in the work environment and relevant objective characteristics of the person (see Figure I.1). Most of the data we have presented, however, consists of subjective verbal responses to a questionnaire. Certainly before applying our findings, it is imperative to understand the extent to which these subjective reports of job stresses correspond to objective stresses.

To examine the relation of objective to subjective stress, we looked at the association between occupation, our only objective environmental variable, and subjective stress (see Chapter VI). The research team chose each of the 23 occupations with the specific aim of maximizing the number of objectively stressful variables and the range on each variable. One indication of our success is the fact that there were significant occupational differences on every one of our measures of stress. More important, those occupations chosen to be high and low on a dimension of objective stress tended to show the same positions on the correponding subjective dimension of stress. For example, air traffic controllers, as expected, were higher than any other occupation on the reported amount of concentration required on their job whereas assembly line workers were very low on this variable. Unfortunately there is no way of quantifying this type of correspondence, but the interested readers can check their expectations for occupational differences against the findings in Appendix A and in the briefer summary in Table VI.2 (page 92).

There were also large occupational differences, in the expected direction, in occupational strains. For example, unskilled blue-collar workers were high on boredom while professionals were low on this variable (see Table VI.1, page 91).

Self-reports of cardiovascular disease and of respiratory disease differed significantly across occupations, and we were not surprised to find that professors were lowest on both since they tend to be low on stresses. In summary, the correlations of mean stress with mean strain across occupations tended to parallel the correlations across individuals.

In all of these correlations of verbally reported stress and strain, we must beware of contamination between the independent and the dependent variable. The men on the assembly line who told us that their jobs were simple and repetitive may have been telling us the same thing over again, in part, when they told us that their jobs were also boring.

Such contamination is not a problem when subjective stress is related to an objectively measured strain such as blood pressure. We find in this study that scientists, compared to six other occupations, had the lowest level of systolic blood pressure (see Figure VI.1, page 91). Although other physiological variables were significantly related to job stresses, we think that the findings regarding blood pressure may be reliable. They replicate and extend previous results on occupational differences in blood pressure (Caplan, 1972).

The most convincing evidence that our subjective measures reflect objective realities is the analysis designed to discover whether the effects of objective occupation on strains can be accounted for by the intervening subjective variables measured in our questionnaire. The results were decisive (see Table VI.3, page 97). We explained from 62 to 79 per cent of the occupation-related variance in the job-related affects (Job Dissatisfaction, Workload Dissatisfaction, and Boredom) and from 25 to 60 per cent of the occupation-related variance in the general affects (Anxiety, Depression, Irritation) and Somatic Complaints by means of the subjective measures of stress, the desired levels of stress, and person–environment fit. Objective occupation explained only from 2 to 6 per cent of variance in the strains beyond that accounted for by subjective measures. We conclude that our subjective measures of stress reflect corresponding objective dimensions of stress and that very little objective job stress is not caught in our measures.

Conditioners of the Effects of Stress on Strain

Conditioning variables may modify the effects of stress on strain. Therefore a number of possible conditioning variables were examined in Chapter V. Only one important effect was discovered. It was found that the effect of Quantitative Workload Fit on job related strains and on general affects were conditioned by fit on complexity: men who had excessive job complexity were more strained as a result of too much workload than were men with good fit on complexity (see Figure V.2, page 82). Given this finding, we would expect that occupations which have poor fit on complexity would also show a strong effect of too much workload whereas much less effect would appear in occupations with good fit on complexity. This prediction was confirmed. Machine-paced assembly line workers had *less* complex work than they wanted, and, as predicted, excessive workload was strongly related to overall psychological strain among them. Train dispatchers, had *more complex* work than they wanted, and excessive workload was also strongly related to overall strain among them. By contrast, white-collar supervisors had good fit on job complexity, and there was no relation between fit on workload and strain (see Figure V.3, page 84). These results suggest a more general principle: *Whenever the effect of a specific stress on a specific strain is conditioned by a third variable, then occupations which differ in the level of this third variable will also differ in the effect of this specific stress on the specific strain.*

Multivariate Analyses

For the testing of theory, and equally for the practical application of the findings, it is crucial to be able to distinguish between causal relations and mere correlations. Our cross-sectional design is not the best for this purpose. The ideal design is an experimental one in which one manipulates one experimental variable, while holding constant all other variables, and observes the resulting changes in the dependent variables. The closest we could come to this ideal in the

Table VII.1   Summary of significant independent predictors of psychological strains as reported in Chapters III and IV

Predictors	Strain (Chapter III)	Strain (Chapter IV)
A. Job Complexity Poor Fit	Job Dissatisfaction, Boredom	Workload Dissatisfaction Anxiety, Boredom
Underutilization of Abilities	Job Dissatisfaction, Boredom	Job Dissatisfaction, Boredom Depression, Workload Dissatisfaction
B. Workload Excess	Workload Dissatisfaction	Irritation
Overtime Fit	Workload Dissatisfaction	
Unwanted Overtime		Workload Dissatisfaction, Anxiety
Workload-Quinn	Workload Dissatisfaction Anxiety	
Responsibility Poor Fit		Boredom
C. Social Support from Others at Work	Depression (–), Irritation (–)	Depression (–), Anxiety (–), Irritation (–), Job Dissatisfaction (–)
D. Participation		Job Dissatisfaction (–)
E. Job Future Ambiguity	Job Dissatisfaction, Depression	Job Dissatisfaction
F. Role Conflict	Irritation	
G. Deny Bad Self	Anxiety (–), Irritation (–), Somatic Complaints (–)	

Note   (–) indicates a negative relationship with the predictor.

current study was to examine the correlation of one presumed independent variable with dependent variables while holding constant many other potentially confounding and intervening variables by means of statistical analysis. This we have done in a series of multiple regression analyses in Chapter III and again in Chapter IV. These analyses were undertaken for somewhat different specific purposes. Chapter III focused on the explanatory power of P–E fit whereas Chapter IV examined the extent to which the most powerful predictors of strain accounted for its variance in multivariate analyses. As a result, the methods used in each chapter were somewhat different. Now we will compare the results of these two series of multiple regression analyses in order to see the extent to which they yielded similar or different conclusions.

Table VII.1 presents the results of these two series of multiple regression analyses, organized in terms of groups of related predictor variables; for example, section A of column 1 deals with the complexity of work in relation to the abilities and needs of the worker, while section B deals with various measures of quantitative workload. The middle column presents the psychological strains which were related to these predictors in the analyses of Chapter III and the right-hand column presents the parallel strains from the analyses in Chapter IV.

The first thing to note about the findings in Table VII.1 is the relatively good agreement between the two sets of multiple regression analyses (columns 2 and 3). Where differences do occur in the two sets of analyses, they are for methodological reasons concerning the criteria for allowing variables to be entered into the multivariate analyses as predictors.[9] Considering the differences in the analytical approaches described in Chapters III and IV, the two sets of results are remarkably and impressively consistent. Now let us summarize the conclusions.

Independent effects found. In the multivariate analyses, we controlled statistically for a large number of (up to 57) potentially confounding variables. As a result, the findings in Table VII.1 are unlikely to be seriously confounded with one another. They are likely to represent *direct* relations as opposed to *indirect* effects mediated by some intervening variables.

---

9. (1) In Chapter III, the variables which were allowed to enter the regression analyses were inter-correlated with one another and conceptually overlapping. This was not the case in the analyses in Chapter IV. As a result, such overlapping variables (such as the measures of social support from supervisor, others at work, and wife, friends, and relatives) often cancelled each other out. Thus Social Support from Others at Work predicted to Job Dissatisfaction and Anxiety in Chapter IV but had no such effect in Chapter III.

(2) In Chapter III, variables were allowed to enter the multivariate analyses if they had significant main effects regardless of the strength of those main effects. In Chapter IV, predictors generally had to have main effects that accounted for 9 per cent of the variance in the dependent variable ($p < .001$). As a result, the role of variables, such as Deny Bad Self, which were examined in Chapter III, were not examined in Chapter IV.

(3) In Chapter III, a predictor that failed to show significant effects in *four* cross-validations of the multivariate analyses was not reported as a significant predictor — a very stringent criterion. In Chapter III, no such cross-validation criterion was set although, as mentioned in the above point, higher levels of significance were required for entry into the analyses.

Qualitative versus quantitative demands of work. Among the effects in Table VII.1, *qualitative* demands of the job represent one of the strongest sets of independent predictors of psychological strain (see section A, column 1). These qualitative demands deal with person–environment misfit with regard to aspects of the job such as complexity and responsibility for persons. They had significant, independent effects on all of the job-related affects: Job Dissatisfaction, Workload Dissatisfaction, and especially Boredom.

Quantitative aspects of the work, as measured by Quantitative Workload, also is an important source of stress (section B, Table VII.1). It affected Workload Dissatisfaction primarily although it also had some effects on Anxiety, Depression and Irritation.

Social support from others at work influences primarily the general affects. These affects are Irritation, Anxiety, and especially Depression. It is possible that the effect may be bi-directional. Depression may cause people to under-report the support they receive. Thus a supervisor who is not willing to listen to the ideas or problems of an employee is likely to produce depression in the employee, which in turn causes the latter to underestimate the small amount of support he or she actually does receive. A likely outcome is misunderstanding and disturbed interpersonal relations.

Participation, an important source of P–E fit. As predicted, Participation showed a negative relation to Job Dissatisfaction in the multiple regression analyses of Chapter IV. This relation, however, did not occur in the analyses of Chapter III where many more variables were controlled. In view of the widespread effects of participation demonstrated in previous research (French and Caplan, 1972), it was even more surprising that no other strains were related to Participation in Table IV.1 (page 66). The explanation of this surprising paucity of effects is revealed in the special analyses (see Figure IV.3, page 64) which showed that the effects of Participation on strain are primarily *indirect* via their effect in improving person–environment fit. Therefore, controlling statistically on these direct effects of person–environment fit greatly reduces the *indirect* effects of Participation.

Role ambiguity and role conflict. Kahn *et al.* (1964) and others have found substantial effects of both Role Ambiguity and Role Conflict on psychological strains. These results were not replicated in this study. Table VII.1 shows no *independent* effects of Role Ambiguity and only one effect of Role Conflict (on Irritation).

Why do these results differ from the earlier findings? In part, our multivariate analyses controlled on two confounding social support variables (from one's supervisor and from others at work) which were correlated with both Role Conflict and Role Ambiguity in our data. These controls were not used in the previous studies. A second reason for the few correlates of Role Conflict is suggested by Sales's (1969) reanalysis of the Kahn *et al.* (1964) data. Sales found that three items measuring role overload could be extracted from the Kahn *et al.* measure of Role Conflict. This measure of role overload correlated .60 with Kahn's measure of Job-Related Tension and negatively with several measures of social support from others in the work setting. It is likely that we found few *independent* effects of role conflict because we controlled on both workload and social support whereas these controls were not used in most of these previous studies of role conflict.

The role of denial. Deny Bad Self was negatively related to Anxiety, Irritation, and Somatic Complaints. Probably this is due partly to a conscious tendency to under-report socially undesirable affects and partly to a less conscious form of denial. The latter defense mechanism has been hypothesized to be an important way of adjusting to stress (see Figure I.1, page 3).

Job Future Ambiguity. Assessing in part the threat of unemployment, this appears positively related to both Job Dissatisfaction and to Depression.

P–E fit versus P and E. Most of the strongest predictors of psychological strain in Table VII.1 were measures of person–environment fit. Overall, these measures of fit accounted for about twice as much variance in strain as did the additive effects of their component measures of the environment and of the person. This result provides strong support for the P–E fit theory presented in Chapter III. Furthermore, the various forms of misfit such as too much of an environmental stress, too little of an environmental supply, or both at once, had the expected effects in our data.

Which particular measure of fit has the greatest effect? This depends on the distribution of the fit scores. For example, excess income cannot have effects in our sample because no one reports excess income. This dependence of the type of misfit on the type of distribution of scores should remind us that even for a single dimension, such as Job Complexity, the distribution of scores, and therefore the type of misfit, will vary across occupations which differ in complexity. In a simple repetitive job, such as an assembly line, the workers will suffer boredom induced by too little complexity, but no such effects will be found in the complex job of the family physician.

One question to be solved in future research involves the extent to which E and P have independent effects beyond the effects of fit. Appendix C shows that the effects of P and E were eliminated in multiple regression analyses partly because our measures of P and E had some overlap with our measures of fit. This is a methodological problem. When a person reports E, such as 'how much workload is there?', the rating scale may partly elicit a measure of fit ('very little', 'very much', and so on). Until rating scales are developed for E and P that are conceptually free of the concept of fit, it will be difficult to assess the extent to which E and P have effects that are independent of the effects of fit. Until that time, the reader is cautioned against concluding that there are no effects of E and P. There do appear to be effects, but they tend to be obscured in the multivariate analyses by the effects of fit, the most powerful of the predictors of strain.

The principle of conceptual relevance. A general hypothesis of this study states that the magnitude of the effect of a psychological stress on a psychological strain will increase with increasing *relevance* of the dependent variable to the independent variable. This hypothesis is supported in Table VII.1 and in several other analyses in preceding chapters. For example Workload Excess affected Workload Dissatisfaction but it had no independent effects on general Job Dissatisfaction nor on the even less relevant affects such as Depression and Anxiety. In general we may conceive of the *relevance* of the dependent variable as the degree to which it is commensurate with the independent variable in the sense that it can be measured with the same units of measurement. For example, maximum relevance to the job demand for typing speed (measured in words per minute) would be typing performance (measured in words per minute).

The importance of relevance is further demonstrated in a comparison of the strong findings on P–E fit with the weak findings on conditioning variables in Chapter V. Both analyses deal with interaction effects, but P–E fit yielded very strong findings (see page 107) because the P variables were constructed to be maximally relevant to (that is, commensurate with) the E variables, whereas the less relevant interaction effects in Chapter V were weak indeed because person variables like Type A were not studied with regard to commensurate environmental variables.

Physiological strains as outcomes. Multiple regression analyses of psychological stresses in relation to physiological strains (see Chapter III) revealed relations that were weak ($r < .15$), inconsistent and often contrary to theoretical predictions. However, we cannot conclude from these findings that psychological variables do not affect physiological strains. There are too many methodological reasons why our findings are weaker than the results of laboratory experiments. Our selection of physiological variables was constrained by costs and by the requirements for simplicity of collection nationwide. Therefore, many physiological variables were omitted, there was no possiblity to study acute stresses, and we could not compare measures before and after a stressor and thus control on many individual differences in personality or in heredity. Perhaps most important, many of our variables, such as heart rate and blood pressure, are responsive to momentary stresses which may mask the effects of the chronic stresses which are the focus of this study.

Implications for Further Research

It is not surprising that a project such as this one should raise more questions than it answers. Not all of the questions for future research will be discussed here. Instead, we will select a few problems which are especially important for theory or application or both. The first few problems are concerned with P–E fit theory.

Need for Measures of Objective Fit

As we have noted above, our theory of person–environment fit emphasizes objective job stresses as a root cause of stress, yet we have few data on objective stresses. Future research should fill this gap by developing measures of objective stresses which are exactly commensurate with our subjective measures of stress. This would provide more knowledge about the role of objective stress. At the same time, the discrepancy between an objective stress and a commensurate subjective stress would yield a measure of 'contact with reality' (see page 4 and Figure I.1) which would open the way for a better understanding of the role of defenses such as denial in adjusting to stress.

The Principle of Relevance

This principle (see page 64) should be extended as a way of increasing the power of P–E fit theory to explain strains. This means that *three* variables must be measured on the same commensurate dimension: (a) the environmental variable, (b) the person variable, (c) the dependent variable. This triple commensurability could be achieved by starting at any point. For example, starting with our measure of fit on job complexity, we could improve the percentage of variance accounted for by developing a more relevant dependent variable than our measure of boredom; for example, we could develop a measure of 'satisfaction with job complexity' where job complexity was specified in exactly the same way as in the measure of fit. Another starting point might be a known relationship between a person variable such as 'time urgency' (Friedman and Rosenman, 1974) and physiological strains such as high blood pressure and a high cholesterol level. In this case it would be necessary to develop measures of a 'time-urgent job environment' and a commensurate dependent variable such as satisfaction with the deadlines in the job.

Operationalizing the Distinction between the Two Main Types of Fit

The theory distinguishes between job demands and abilities and the fit between the person's needs and the environmental supplies for these needs. In this project our measures do not distinguish these two clearly enough for us to determine whether they produce, as expected, different strains. This problem deserves further study.

Experiments on the Effects of Participation on P–E Fit

Our findings suggest that most of the effects of participation on strain take place indirectly via their effects on person–environment fit. Other studies, many of them field experiments, show very widespread effects of participation (French and Caplan, 1972), but it is not known whether person–environment fit is part of the mechanism for these effects. Furthermore, many of the experiments on participation have demonstrated effects on productivity as well as on good working relations with others, good affective states, and good health, among other factors (French, 1975). The time has come when we need field experiments to test the intervening role of person–environment fit in the effects of participation on strain and on productivity. Such studies will be important not only in clarifying theory but also in determining whether the costly efforts to reduce strains are balanced by tangible benefits such as increased productivity and lowered costs. Knowledge of these costs and benefits will greatly influence the efforts to utilize participation to improve person–environment fit and thereby well-being.

Studies of Other Causal Relations among Stresses

Other job stresses besides low Participation (for example, Job Complexity-E, Responsibility for Persons-E) do not appear as independent predictors of strain in the multiple regression analyses summarized in Table VII.1, yet in previous research they are correlated with other stresses and with strains (Caplan et al., 1980, Appendix G). It may be that these stresses affect strains indirectly by means of causal influences on other stresses. If so, they might provide means for controlling these other stresses. Research on these causal relations among stresses is required before such practical applications can be developed.

Generalizability of the Research Methods

The success of our research methods in accounting for occupational differences in stress and strain suggests that they might fruitfully be applied in other cases where it is already known that there are large occupational differences in diseases and/or in physiological risk factors, such as the occupational differences in blood pressure found in this study. If there is reason to believe that such known differences may be related to stress, then such research has a good chance of contributing to our understanding of stress and health. For example, the research might identify the dimensions of person–environment fit that are causing the health problems and thus point to the kind of changes which would prevent these problems.

Generalizing to Female Populations

As noted in Chapter II, there were too few females in the occupations studied to obtain an adequate sample for analyses. One study, however, has used many of our measures of job demands, fit, and strain and was able to compare the results for male and female respondents. Singer (1975) surveyed a random sample of male and female employees involved in enforcement in a federal agency, an occupation not in our sample. In that study, there were 'moderate, positive relationships between most P–E fit discrepancies and strain *for both sexes*' (p.199). There were only a few exceptions where male and female data showed different results. Females perceived more Role Ambiguity than males and the relationship between Role Ambiguity and Job-Related Tension was much stronger for females than for males. Singer found that females in his study had higher levels of life stress than males and, furthermore, that their life stress scores accounted for most of the variance in job-related strains.

The mean levels of psychological strain (Anxiety, Depression, Dissatisfaction, and so forth) were similar for males and females. This is in contrast, however, with national sample data (Gurin, Veroff and Feld, 1960) which found that women reported more Somatic Complaints whereas men reported more Job-Related Tension. The national sample findings, however, did not control for occupational differences in the sexes.

Overall, the similarities of the male and female findings in the Singer study of enforcement personnel are particularly striking given that there were demographic differences between the males and females in age, marital status, and education. Females were younger (and so probably had less tenure in the occupation), were more likely to have been divorced or never married, and had less formal education, but the same percentage of males and females held supervisory positions.

Additional studies are needed on females in other occupations in order to determine the extent to which the results of our study of job demands and worker health apply to females in a variety of occupations. The results of the Singer study suggest that, in the main, the relationships that we have reported would be likely to hold for females as well.

Implications for Possible Application

Ethical Issues of Application

Inevitably the application of research findings raises ethical questions which cannot be answered by scientific research. However, research can help to clarify these problems. In this project there is one pervasive ethical question which must be faced in the beginning: should one attempt to change subjective variables without any corresponding changes in objective variables? For example, should one attempt to change subjective stresses and perceived hazards in the workplace without doing anything about the objective stresses and hazards? Or should one attempt to change a person's perception of his or her own abilities without any change in the objective abilities? These questions imply the further question: should one attempt to change the subjective fit of a person without any corresponding change in the objective fit?

The answers to these questions depend very much on the initial relations between the objective and subjective variables and the direction of the proposed change. For example, if a person underestimates an environmental stress or threat, then an increase in the subjective threat will make the person both feel more threatened and be more realistic, whereas a decrease in the perceived threat will make the person feel less threatened and also cause the person to be less realistic. Accepting for the moment, the common assumption that improved contact with reality is an indicator of better mental health, we then sometimes face the dilemma that an improvement in subjective fit may be counterbalanced by a decrease in contact with reality. In other cases, however, a given change will improve both fit and contact with reality. For example, consider a person who reports less opportunity for promotion than the person desires. Assume that more opportunity than the person perceives *is* available and that the person is then told about such opportunity. As a result, the person will have better contact with reality and better P–E fit on opportunity for promotion. A further complication arises from qualifications about the desirability of completely accurate perceptions of environmental stresses and of the self. It may well be, as Lazarus (1979) has argued that a temporary denial of an objective stress may reduce the level of anxiety so that a person can better cope with the objective situation. These same problems exist for discrepancies between the objective person and the subjective person and for discrepancies between objective fit and subjective fit.

We do not believe these ethical dilemmas can be reasonably resolved until one has better information about the relative merits of coping and defense under specific conditions. Accordingly, the following discussion about possible applications of our findings will be limited to the discussion of objective changes or to subjective changes which do not decrease the person's accuracy in perceiving the environment and the self. At the same time we recognize that persons do adjust to stress with defensive distortions and that very often these choices may be best for the long-run health and adjustment of the person. However, it is one thing for the *person* to make these decisions, but it is quite a different thing for *others* to make these decisions without the participation of the person. So, our discussion of intervention by the organization or by third persons will concentrate on those applications which do not increase distortions.

Dealing with Occupational Differences in Stress—
Which Stresses and Strains can be Changed?

Although there are large occupational differences in stress and strain, it is unreasonable to attempt to eliminate high stress and strain by eliminating these stressful occupations. Instead, we must deal with the dimensions of stress (including both job demands and misfit) which are causing the strains. The finding that an average of 55 per cent of the variance in strains attributable to occupation can be explained by our measures of the job environment and the person (see Table VI.3, page 97) suggests that much of the problem of occupational differences in strain can be approached by dealing with job-stress variables.

Dealing with main effects of stress. Most of the important predictors of strain, as summarized in Table VII.1, are variables that can be manipulated or changed. Variables in section A of Table VII.1 on the complexity of the work have often been changed in programs of job enrichment. The quantitative workload in the job (section B) is easily changed in either direction in the abstract; but in practice it is always an issue of social conflict between managers and employees. Participation, which has some of the most widespread effects (Tannenbaum, 1968), has been experimentally changed by means of laws which mandate worker participation (Kavcic, Rosnec, Vianello, *et al.*, 1974) and by planned interventions in specific organizations (Campbell, 1974; Coch and French, 1948; French, Israel, and Aas, 1960; French, Kay and Meyer, 1966).

It is clear that organizations vary greatly in the degree to which they provide *social support* for the members working in the organization. Few experiments, however, have actually varied this parameter of stress in a work organization. Nevertheless, it is clear in principle that changes in supervisory practices and changes in the cohesiveness of work groups can be carried out to increase social support.

Of all the job stresses that we have studied, perhaps the most difficult to control is Job Future Ambiguity. Such ambiguity depends heavily on the state of the economy and is not fully under the control of a single organization. A detailed study of a plant closing shows that the anticipation of the event as well as the occurrence has serious effects on mental and physical health (Cobb and Kasl, 1977; Kasl and Cobb, 1979).

Dealing with interaction effects of stress. In the above summary of results we have noted that occupations differ on the *effects* of stress on strain as well as on the level of stress or strain. Our findings on person–environment fit with respect to Job Complexity support the general principle that whenever the effects of a stress on a strain are conditioned by a third variable, then occupations which differ in the level of this third variable will also differ from one another in the effect of the stress on the strain. We see evidence for this in the study. Fit on Job Complexity conditioned the relationship between fit on Workload and Strain. And occupations with different levels of fit on Job Complexity differed in the effects of misfit on workload and strain.

Two other interaction effects are prominent in our findings and have implications for occupational difference in stress-change strategies: (a) the powerful effects of person–environment fit represent an interaction of its component variables, and (b) social support was found to have widespread effects in buffering the effects of stress on strain. These latter effects were only briefly mentioned on page 106, because a recent article deals extensively with this topic (LaRocco, *et al.*, 1980). In these cases, occupations that differ in P (needs, traits) may differ in how commensurate stresses affect the strains. Similarly, occupations with low social support may show stronger effects of job demands on strain than occupations with high social support. In the latter case, one might focus on reducing stress in occupations with low social support because those occupations lack the buffering resources of support (Cobb, 1976; LaRocco *et al.*, 1980).

We conclude then that whenever the occupations vary in the level of one or more of these known conditioning variables, any effects of stress on strain are likely to vary from one occupation to another, and such variation will be predictable. Accordingly, it is essential to obtain a good diagnosis of the relations among these variables in specific occupations before planning interventions aimed at reducing stress and strain. This can be done by (a) measuring stresses, strains, and conditioning variables, as we have done; and (b) then conducting analyses, similar to the ones reported above, to determine how occupations differ in the effects of a given stress on a given strain. Such information will permit a pinpointed application of the findings in those occupations and with respect to those stresses where the effects of stress are greatest.

There is one special advantage of discovering interaction effects: they present two clear options for interventions to reduce strain. For example, if one knows that Complexity-Fit conditions the effects of Workload Fit on strains such as Dissatisfaction with Workload, then there are two ways to reduce the strain. Either one can reduce the quantitative overload or alternatively one can improve the qualitative nature of the work. The particular choice between these options may well depend upon costs and other constraints which impact more strongly on one option than on the other. To given another example, in a given occupation, a careful diagnosis may reveal that a specific psychological stress is producing widespread strains but it is impossible or too costly to reduce the stress; in this case, these effects of stress on strains may be effectively reduced or prevented by increasing the amount of social support provided for the employees.

Improving Person–Environment Fit

Any intervention designed to improve person–environment fit *must be individualized*. A standard job enlargement program might improve the fit of some workers while worsening the fit of others. For example, for the policemen in our sample, such a program would produce worse fit in the majority while improving fit in a large minority. Therefore, *individual* data on person–environment fit are required for planning a successful program. Our questionnaires can provide these data for a wide spectrum of jobs, but some dimensions may need to be added for special jobs: for example, one special stress common for policemen is hostile reactions from members of the public (French, 1975).

There are several ways of improving fit in an individualized program: by changing job stresses, by changing abilities of the person, by increasing supplies for needs, by changing the person's goals, and by a process of matching persons with jobs. We will discuss each one of these methods starting with the last one.

Personnel selection procedures. Given a set of jobs and a set of persons to occupy the jobs it is possible to improve person–environment fit, without any changes in the job environment or any changes in the person, simply by the optimal matching of persons with jobs. Traditional personnel selection procedures have long been involved with practices designed to match the abilities of the employees with the demands of the job, but studies of the validity of selection tests and procedures have usually revealed a low level of success, partly because these procedures have paid too little attention to the ability to withstand job stresses. In principle, given a variety of jobs and a variety of applicants, the stresses and supplies of each job can be assessed and matched with the abilities and motives of the applicants in such a way as to optimize the fit between the person and the job. In practice, this often fails because it is based on static criteria in a changing world. Both the job demands and the needs and abilities of the person will change. It is especially difficult to predict in advance an enduring state of person–environment fit.

Reducing job stress. Improving the person–environment fit of an employee by changing job stresses may run into important opposing forces. For example, the change may involve changes in technology which are extremely expensive. In addition, the individualized changes in specific job demands upsets the formal bureaucratic structure of jobs and may result in many difficulties in the control and coordination of jobs when each individual is doing his or her own thing. Furthermore, there may be serious problems of inequity of pay and other reward when jobs are individualized rather than standardized. For example, if standardized workloads are adjusted so that some individuals are expected to do more than others, this will be perceived as inequitable unless there are some corresponding adjustments in pay. Despite these barriers to an individualized change in job stresses in order to improve person–environment fit, it is well worth the effort. We have found that person–environment fit can be twice as important as the job stresses plus the effects of personal abilities and goals in determining job strains. Furthermore, when we consider the possibly conflicting demands for productivity, for increased efficiency, and for profits, it seems likely that improved person–environment fit may often lead to improved efficiency. For example, the stress on the person due to the underutilization of the employee's highest skills means that the organization is not getting its full return from the money invested in the person's salary. Secondly, at the opposite end of the dimension of qualitative misfit, when the quality demands of the job exceed the corresponding abilities of the person, we can be sure that the organization's standard for quality and performance will not be met.

We have noted in the summary of results that the variable of low participation produces far fewer strains than we might expect on the basis of previous research, including experimental research. Our analysis shows that this is probably due to the fact that participation has its major effects on other stresses rather than directly on other strains. In short, the effect of participation is to reduce many other stresses including person–environment misfit, and these reduced misfits in turn reduce many psychological strains.

When we consider both the great advantages of individualized efforts to improve person–environment fit and the opposing forces in a highly technological and bureaucratic structure it suggests the great importance of *flexibility* in the role structure of any work organization. Work organizations must show flexibility of role structures, policies, and procedures, if they are to maximize the adjustment of their members to sometimes unavoidable stresses of organizational membership.

It is possible that the necessary flexibility in the organizational structuring of roles required for people to make adjustments in P–E fit will occur primarily in organizations operating in turbulent environments (see, for example, Bennis and Slater, 1968; Burns and Stalker, 1961; and Lawrence and Lorsch, 1967). Such organizations may have an incentive to survive by continually allowing their employees to make adjustments in role relations with one another or by taking steps to stabilize the environment. Organizations in more stable environments may better survive by developing a stable set of role relationships and persons to fit those relationships via processes of personnel selection and attrition. Research on the relationship between environmental turbulence and optimal flexibility of organizational structure paints an unclear picture of whether such flexibility is optimal for organizational survival (Katz and Kahn, 1978). Until that question is answered more adequately, it is unlikely that organizations will seek flexible role structures as a mode for providing personnel with opportunities to improve person–environment fit.

Increasing employee abilities. Improving person–environment fit by changing the abilities of the employees is again a very common goal for vocational education and for on-the-job training. Again, it is unfortunate that training has too often been devoted only to increasing fit on quantitative workload while too little attention has been devoted to using training programs to improve the fit of the employee to his or her job on dimensions such as role ambiguity and responsibility for others.

Even before the usual age of becoming an employee, the processes of anticipatory socialization into occupational roles are widespread in our society. In vocational schools and vocational courses in high school, the education is intended to fit a student for some specific occupation, such as typist, automobile mechanic or carpenter. Higher education provides the knowledge, skills and attitudes for one of the professions or for other specialized occupations. In general, such anticipatory socialization provides a standardized curriculum which pays little attention to psychological job stresses and to instruction on how to cope with them. Obviously, this need not be the case, and as knowledge of the processes of adjustment to job stress increases, this knowledge should be applied in occupational education and training.

Dynamic adjustment. Once a person is hired into a job, there is an opportunity for continuous improvements in person–environment fit. The orientation and on-the-job training can be more specific and more individualized to match the characteristics of the employee with the job. On-the-job training, at least informally, is more apt to deal with problems of adaptation to job stresses. Although new employees may experience ambiguity and overload, our results suggest that attention should also be paid to the person–environment fit of those who are overqualified for the jobs. We find that underutilization of abilities is still one of the most common job stresses, and it is also one of the most wasteful practices of work organizations in the efficient accomplishment of their major goal of production. As employees learn more about the jobs in their organizations the threat of underutilization grows and the need for continuous job enrichment and/or promotion increases.

The employee's versus the employer's perspective on P–E fit. So far, our discussion of ways of improving person–environment fit has taken the viewpoint primarily of the manager or of the organization which is attempting some intervention. Now, we change this orientation and ask: what is the implication of our research for actions and adjustive processes which might be taken by the employee who is subjected to job stresses? This change in perspective is necessary partly because the two perspectives often involve a conflict of interests.

It is to the interest of employees to remove the misfit between their actual wages and their aspired wages by increasing the actual wages, but it is to the interest of the employer to reduce this gap by reducing the aspirations of the workers. Besides these differences in the interests of the two parties, there are also differences in the available options they are likely to exercise as modes of dealing with fit. The employer may have the options of increasing an employee's fit by altering the objective nature of the supplies (lighting, secretarial help, and so on), rewards (pay, fringe benefits, etc.), opportunities for skill and ability development of the employees, and reduction of job demands. The employee, on the other hand, has some *additional* options that include leaving the job as well as the use of psychological defenses that distort the perception of fit (for example, denying a lack of abilities or an inadequacy of rewards). Although people do attempt to influence one another's subjective perceptions of fit in daily interaction (for example, you attempt to convince someone to overcome an irrational fear), we noted earlier in this chapter that attempts to improve P–E fit by altering the person's subjective perception of fit without making it commensurate with objective fit can raise ethical problems. These differences in options deserve serious consideration in the design of interventions.

A Look into the Future

What promise does this research hold for the future quality of working life? The answer must depend on the evaluation of the results obtained in this particular project, on the prospects for future research that can be built on these findings, and on how well this type of research on occupational stress can provide successful practical answers for improving the quality of working life.

On the optimistic side, we believe we have identified and measured a substantial proportion of the important *subjective* job stresses that produce occupational differences in psychological strains such as job dissatisfaction, boredom, anxiety, and depression. This is an important first step that is essential before one can devise methods for controlling these stresses and hence the strains. Furthermore, our significant findings on person-environment fit point the way to future studies that can surely extend our knowledge of subjective stress and strain in work life.

There is less reason for optimism when it comes to more *objective* indicators of stress and strain. This project was limited to subjective measures of stress, and it is only a strong inference that these measures represent corresponding objective stresses. Future research must develop commensurate measure of objective stresses and of objective person-environment fit, a task that will require some years. Objective measures of the quality of life, such as physiological indices and health measures, are already developed; but in this project we did not find them to be meaningfully related to subjective stresses. There were just enough statistically significant findings to encourage further research on health variables. Such research should examine simultaneously the influence of psychological stress and the influence of physical factors. The continuing debate on one of these physical factors, whether a low cholesterol diet improves health, should warn us not to expect quick practical results from such research.

Health versus production. This project, like most research on stress, has focused on individual outcomes rather than organizational outcomes such as productivity and the viability of the organization. Both types of outcomes play a part in the conflicts between management and labor, but the relation between them is not well understood. Studies of the relationship of job demands and worker health to productivity and economic outcomes are needed to better understand the extent to which goals mutually acceptable to employee and employer may be reachable by paths mutually acceptable to both parties.

References

Adams, J. S. Inequity in social exchange. In L. Berkowitz (ed.), *Advances in Social Psychology*. New York: Academic Press, 1965.

Althauser, R. P. Multicollinearity and non-additive regression models. In H. M. Blalock, Jr. (ed.), *Causal models in the social sciences*. Chicago: Aldine, 1971.

Andrews, F. M., Morgan, J. M., Sonquist, J. A. and Klem, L. *Multiple Classification Analysis: A Report on a Computer Program for Multiple Regression Using Categorical Predictors*. Ann Arbor, Michigan: Institute for Social Research, 1973.

Barnowe, J. T., Mangione, T. W., and Quinn, R. P. Quality of employment indicators, occupational classifications, and demographic characteristics as predictors of job satisfaction. In R. P. Quinn and T. W. Mangione (eds.) *The 1969–70 Survey of Working Conditions: Chronicles of an Unfinished Enterprise*. Ann Arbor: Survey Research Center, 1973, pp.385–392.

Bennis, W. G. and Slater, P. E. *The Temporary Society*. New York: Harper, 1968.

Bentler, P. M., Jackson, D. N. and Messick, S. Identification of content and style: A two-dimensional interpretation of acquiescence. *Psychological Bulletin*, 1971, **76**, 186–204.

Blauner, R. *Alienation and Freedom; The Factory Worker and his Industry*. Chicago: University of Chicago Press, 1964.

Block, J. *The Challenge of Response Sets*. New York: Appleton, 1965.

Box, G. E. P. A general distribution theory for a class of likelihood criteria. *Biometrica*, 1949, **36**, 317–346.

Burns, T. and Stalker, G. M. *The Management of Innovation*. London: Tavistock, 1961.

Campbell, A., Converse, P. E. and Rodgers, W. L. *The Quality of American Life*. New York: Russell Sage Foundation, 1976.

Campbell, D. B. A program to reduce coronary heart disease risk by altering job stresses (doctoral dissertation, University of Michigan, 1973). *Dissertation Abstracts International*, 1974, **35**, 564B. (University Microfilms No. 74-15681).

Caplan, R. D. Organizational stress and individual strain: a social-psychological study of risk factors in coronary heart disease among administrators, engineers, and scientists (doctoral dissertation, University of Michigan, 1971). *Dissertation Abstracts International*, 1972, **32**, 6706B–6707B. (University Microfilms No. 72-14822).

Caplan, R. D., Cobb, S., French, J. R. P., Jr., Harrison, R. V. and Pinneau, S. R., Jr. *Job Demands and Worker Health: Main Effects and Occupational Differences*. Ann Arbor, Michigan: Institute for Social Research, 1980.

Caplan, R. D., and Jones, K. W. Effects of work load, role ambiguity, and personality Type A on anxiety, depression, and heart rate. *Journal of Applied Psychology*, 1975, **60**, 713–719.

Cobb, S. *Class A Variables from the Card Sort Test*. (A Study of People Changing

Jobs, Project analysis memo # 12). Ann Arbor, Michigan: University of Michigan, Institute for Social Research, July 24, 1970.

Cobb, S. Physiologic changes in men whose jobs were abolished. *Journal of Psychosomatic Research*, 1974, **18**, 245–258.

Cobb, S. Social support as a moderator of life stress. *Psychosomatic Medicine*, 1976, **38**, 300–314.

Cobb, S. and Kasl, S. V. *Termination: The Consequences of Job Loss*. Cincinnati, Ohio: DHEW (NIOSH) Publication No. 77-224, 1977.

Cobb, S. and Rose, R. M. Hypertension, peptic ulcer and diabetes in air traffic controllers. *Journal of the American Medical Association*, 1973, **224**, 489–492.

Cobb, S., Tomlin, P., Chun, K. T., French, J. R. P., Jr. and Kasl, S. V. B Bias and error in self report of body weight (unpublished manuscript). Ann Arbor, Michigan: Institute for Social Research, 1974.

Coch, L. and French, J. R. P., Jr. Overcoming resistance to change. *Human Relations*, 1948, **1**, 1–21.

Cohen, J. Multiple regression as a general data-analytic system. *Psychological Bulletin*, 1968, **70**, 426–443.

Crowne, D. P. and Marlowe, D. *The Approval Motive*. New York: Wiley, 1964.

Darlington, R. B. Multiple regression in psychological research and practice. *Psychological Bulletin*, 1968, **69**, 161–182.

Dunbar, F. *Mind and Body: Psychosomatic Diagnosis*. New York: Hoeber, 1948.

Dunn, J. P., Brooks, G. W., Mausner, J., Rodnan, G. P. and Cobb, S. Social class gradient of serum uric acid levels in males. *Journal of the American Medical Association*, 1963, **185**, 431–436.

Evans, M. G. Conceptual and operational problems in the measurement of various aspects of job satisfaction. *Journal of Applied Psychology*, 1969, **53**, 93–101.

Florey, C. V. The use and interpretation of ponderal index and other weight–height ratios in epidemiological studies. *Journal of Chronic Diseases*, 1970, **23**, 93–104.

Fox, D. and Guire, K. *Documentation for MIDAS* (Michigan Interactive Data Analysis System) (2nd edition). Ann Arbor, Michigan: Statistical Research Laboratory, University of Michigan, 1976.

French, J. R. P., Jr. Person role fit. *Occupational Mental Health*, 1973, **3**, 15–20.

French, J. R. P., Jr. A comparative look at stress and strain in policemen. In W. H. Kroes and J. J. Hurrell (eds.), *Job Stress and the Police Officer: Identifying Stress Reduction Techniques*. US Department of Health, Education, and Welfare. HEW Publication No. (NIOSH) 76-187, 1975.

French, J. R. P., Jr. and Caplan, R. D. Psychosocial factors in coronary heart disease. *Industrial Medicine*, 1970, **39**, 383–397.

French, J. R. P., Jr. and Caplan, R. D. Organizational stress and individual strain. In A. J. Marrow (ed.), *The Failure of Success*. New York: AMACOM, 1972.

French, J. R. P., Jr., Israel, J. and Aas, D. An experiment in participation in a Norwegian factory. *Human Relations*, 1960, **13**, 3–19.

French, J. R. P., Jr., Kay, E. and Meyer, H. H. Participation and the appraisal system. *Human Relations*, 1966, **19**, 3–20.

French, J. R. P., Jr., Rodgers, W. L. and Cobb, S. Adjustment as person–environment fit. In B. V. Coelho, D. A. Hamburg, and J. E. Adams (eds.). *Coping and Adaptation*. New York: Basic Books, 1974, pp.316–333.

Frenkel-Brunswik, E. Intolerance of ambiguity as an emotional and perceptual personality variable. *Journal of Personality*, 1949, **18**, 108–143.

Friedman, M. and Rosenman, R. H. *Type A Behavior and Your Heart*. New York: Alfred A. Knopf, 1974.

Glass, D. C. *Behavior Patterns, Stress, and Coronary Disease*. Hillsdale, New Jersey: Lawrence Erlbaum Associates, 1977.

Goldberger, A. S. and Jochem, D. B. Note on stepwise least squares. *Journal of the American Statistical Association*, 1961, **56**, 105–110.

Goldbourt, U. and Medalie, J. H. Weight–height indices. *British Journal of Preventive and Social Medicine*, 1974, **28**, 116–126.

Gore, S. The influence of social support and related variables in ameliorating the consequences of job loss (doctoral dissertation, University of Pennsylvania, 1973). *Dissertation Abstracts International*, 1974, **34**, 5330A–5331A. (University Microfilms No. 74-2416).

Gough, H. G. *California Personality Inventory Manual*. Palo Alto, California: Consulting Psychologists Press, 1957.

Gurin, G., Veroff, J. and Feld, S. *Americans View their Mental Health*. New York: Basic Books, 1960.

Hackman, J. R. and Lawler, E. E. Employee reactions to job characteristics. *Journal of Applied Psychology*, 1971, **55**, 259–286.

Harrison, R. V. Job demands and workers health: Person environment misfit (doctoral dissertation, University of Michigan, 1976). 320pp. *Dissertation Abstracts International*, 1976. (University Microfilms # 76–19, 150).

House, J. S. The relationship of intrinsic and extrinsic work motivations to occupational stress and coronary heart disease risk (doctoral dissertation, University of Michigan, 1972). *Dissertation Abstracts International*, 1972, **33**, 2514A. (Universty Microfilms No. 72-29094).

House, J. S. *Occupational Stress and the Mental and Physical Health of Factory Workers*. Ann Arbor, Michigan: Institute for Social Research Report Series, 1980.

Institute for Social Research. *OSIRIS III: An Integrated Collection of Computer Programs for the Management and Analysis of Social Science Data*. Ann Arbor: The University of Michigan, 1973.

Jahoda, M. *Current Concepts of Positive Mental Health*. New York: Basic Books, 1958.

Kahn, R. L. and Quinn, R. P. Role stress: A framework for analysis. In A. McLean (ed.), *Occupational Mental Health*. New York: Rand McNally, 1970.

Kahn, R. L., Wolfe, D. M., Quinn, R. P., Snoek, J. D. and Rosenthal, R. A. *Organizational Stress: Studies in Role Conflict and Ambiguity*. New York: Wiley, 1964.

Kantor, R. *Work and Family in America: A Critical Review and Research Agenda*. New York: Russell Sage (Social Science Frontiers Monograph Series), 1977.

Karasek, R. A., Jr. The impact of the work environment on life outside the job. Doctoral dissertation, Massachusetts Institute of Technology, 1976.

Karasek, R. A., Jr. Job demands, job decision latitude and mental strain: implications for job redesign. *Administrative Science Quarterly*, 1979, **24**, 285–308.

Kasl, S. V. and Cobb, S. Some mental health consequences of plant closing and job loss. In L. A. Ferman and J. P. Gordus (eds.), *Mental Health and the Economy*. Kalamazoo, Michigan: W. E. Upjohn Institute for Employment Research, 1979.

Kasl, S. V., Cobb, S. and Brooks, G. W. Changes in serum uric acid and cholesterol levels in men undergoing job loss. *The Journal of the American Medical Association*, 1968, **206**, 1500–1507.

Kasl, S. V. and French, J. R. P., Jr. The effects of occupational status on physical and mental health. *Journal of Social Issues*, 1962, **18**, 67–89.

Katz, D. and Kahn, R. L. *The Social Psychology of Organizations* (2nd ed.). New York: Wiley, 1978.

Kavcic, B., Rosner, M., Vienello, M., Wieser, G. and Tannenbaum, A. *Hierarchy in Organizations: An International Comparison*. San Francisco: Jossey-Bass, 1974.

Keenan, A. and McBain, G. D. M. Effects of type A behavior, intolerance of ambiguity, and locus of control on the relationship between role stress and work-related outcomes. *Journal of Occupational Psychology*, 1979, **52**, 277–285.

Kessler, R. C. Stress, social status, and psychological distress. *Journal of Health and Social Behavior*, 1979, **20**, 259–272.

Kohn, M. L. *Class and Conformity: A Study in Values*. Homewood, Illinois: Free Press, 1969.

Kornhauser, A. W. *Mental Health of the Industrial Worker: A Detroit Study*. New York: Wiley, 1965.

Kulka, R. A. Person–environment fit in the high school: a validation study (2 vols) (doctoral dissertation, University of Michigan, 1975). *Dissertation Abstracts International*, 1976, **36**, 5352B. (University Microfilms, No. 76-9438).

Langner, T. S. A 22-item screening score of psychiatric symptoms indicating impairment. *Journal of Health and Human Behavior*, 1962, **3**, 269–276.

LaRocco, J. M., House, J. S. and French, J. R. P., Jr. Social support, occupational stress, and health. *Journal of Health and Social Behavior*, 1980, **21**, 202–218.

Lawler, E. E. *Motivation in work organizations*. Belmont, California: Wadsworth Publishing Company, 1973.

Lawrence, P. R., and Lorsch, J. W. *Organization and Environment*. Boston: Harvard Business School, Division of Research, 1967.

Lazarus, R. S. *Psychological Stress and the Coping Process*. New York: McGraw-Hill, 1966.

Lazarus, R. S. Positive denial: The case for not facing reality. *Psychology Today*, **November 1979**, 44–60.

Lazarus, R. S. and Launier, R. Stress-related transactions between person and environment. In L. A. Pervin and M. Lewis (eds.), *Internal and External Determinants of Behavior*. New York: Plenum, 1978.

Levi, L. (ed.). *Stress and Distress in Response to Psychosocial Stimuli*. Oxford: Pergamon Press, 1972.

Lewin, K. *Field Theory in Social Science*. New York: Harper, 1951.

Likert, R. *New Patterns of Management*. New York: McGraw-Hill, 1961.

Lillibridge, J., Jr. *Suggested Personality Indices Based on Factor Analysis of Personality Scale Items*. Ecology of Employment Termination Project analysis memo # 9. Ann Arbor, Michigan: Institute for Social Research, 1970.

Mason, J. W. A review of psychoendocrine research on pituitary and thyroid hormones. *Psychosomatic Medicine*, 1968, **30**, 666–681.

Mason, J. W. Organization of psychoendocrine mechanisms: A review and reconsideration of research. In N. S. Greenfield and R. A. Sternbach (eds.). *Handbook of Psychophysiology*. New York: Holt, Rinehart and Winston, 1973.

Morgan, J. N. Consumer investment expenditures. *American Economic Review*, 1958, **48**, 874–902.

Mueller, E. F. and French, J. R. P., Jr. Uric acid and achievement. *Journal of Personality and Social Psychology*, 1974, **30**, 336–340.

Murray, H. A. *Explorations in Personality*. New York: Oxford University Press, 1938.

Nunnally, J. C. *Psychometric Theory*. New York: McGraw-Hill, 1967.

Obrist, P. A., Black, A. H., Brener, J. and DiCara, L. V. (eds.). *Cardiovascular Psychophysiology: Current Issues in Response Mechanisms, Biofeedback and Methodology*. Chicago: Aldine, 1974.

Osipow, S. H. *Theories of Career Development*. New York: Appleton-Century-Crofts, 1968.

Pervin, L. A. Performance and satisfaction as a function of individual–environment fit. *Psychological Bulletin*, 1968, **69**, 56–68.

Pflanz, M., Rosenstein, E. and Von Uexkull, T. Socio-psychological aspects of peptic ulcer. *Journal of Psychosomatic Research*, 1956, **1**, 68–74.

Pinnea, S. R. *Complementarity and Social Support* (unpublished manuscript). Ann Arbor: Institute for Social Research, University of Michigan, 1972.

Pinneau, S. R. *Satisfaction, Dissatisfaction, and Their Dimensionality* (unpublished document). Ann Arbor, Michigan: Institute for Social Research, 1973.

Pinneau, S. R. Effects of social support on psychological and physiological strains (doctoral dissertation, University of Michigan) Ann Arbor, Michigan: University Microfilms, 1975, No. 76-9491.

Quinn, R., Seashore, S., Kahn, R., Mangione, T., Campbell, D., Staines, G. and McCollough, M. *Survey of Working Conditions: Final Report on Univariate and Bivariate Tables.* (Document No. 2916-0001). Washington, DC: US Government Printing Office, 1971.

Quinn, R. P. and Shepard, L. J. *The 1972–73 Quality of Employment Survey: Descriptive Statistics with Comparison Data from the 1969–70 Survey of Working Conditions.* Ann Arbor: Survey Research Center, 1974.

Reiss, A., Duncan, O., Hatt, P. and North, C. *Occupations and Social Status.* New York: Free Press, 1961.

Rokeach, M. *The Nature of Human Values.* New York: Free Press, 1973.

Rosenberg, M. *The Logic of Survey Analysis.* New York: Basic Books, 1968.

Rosenman, R. H., Friedman, M., Straus, R., Jenkins, C. D., Zyzanski, S. J. and Wurm, M. Coronary heart disease in the Western Collaborative Group Study, a follow-up experience at 4½ years. *Journal of Chronic Diseases,* 1970, **23**, 173–190.

Russek, H. I. Stress, tobacco, and coronary heart disease in North American professional groups. *Journal of the American Medical Association,* 1965, **192**, 189–194.

Sales, S. M. Differences among individuals in affective, behavioral, biochemical, and physiological responses to variations in work load (doctoral dissertation, University of Michigan, 1969). *Dissertation Abstracts International,* 1969, **30**, 2407-B. (University Microfilms No. 69-18098).

Sales, S. M. and House, J. Job dissatisfaction as a possible risk factor in coronary heart disease. *Journal of Chronic Diseases,* 1971, **23**, 861–873.

Scheffler, J., Rice, R. and Kaplan, J. *Occupation Reference File Workbook.* Ann Arbor, Michigan: Institute for Social Research, 1971.

Selye, H. *The Stress of Life.* New York: McGraw-Hill, 1956.

Singer, J. N. Job strain as a function of job and life stresses (unpublished doctoral dissertation). Fort Collins, Colorado: Colorado State University, 1975.

Spicer, J. *Dimensions of Psychological Predisposition to Coronary Heart Disease* (unpublished manuscript). Universty of Otago, New Zealand, 1980.

Spielberger, C. D., Gorsuch, R. L. and Lushene, R. E. *Manual for the State-Trait Anxiety Inventory.* Palo Alto, California: Consulting Psychologists Press, 1970.

Stamler, J., Berkson, D. M., Lindberg, H. A., Miller, W. A., Sevens, E. L., Soyugenc, R., Tokich, T. J., and Stamler, R. Heart rate: an important risk factor for coronary mortality, including sudden death–ten-year experience of the peoples Gas Company Epidemiologic Study (1958–68). Paper presented at the Second International Symposium on Atherosclerosis, Chicago, November, 1969.

Suits, D. B. Use of dummy variables in regression equations. *Journal of the American Statistical Association,* 1957, **52**, 548–551.

Tannenbaum, A. S. *Control in Organizations.* New York: McGraw-Hill, 1968.

Taylor, J. C. and Bowers, D. G. *Survey of Organizations.* Ann Arbor, Michigan: Institute for Social Research, 1972.

Vickers, R. A short measure of the Type A personality (unpublished document). Ann Arbor: Institute for Social Research, University of Michigan, January, 1973.

Vickers, R. Subsetting procedure for the Sales Type A Personality Index: A short measure of the Type A personality. In Caplan, R. D., Cobb, S., French, J. R. P., Jr., Harrison, R. V. and Pinneau, S. R., Jr. *Job Demands and Worker Health: Main Effects and Occupational Differences.* Cincinnati: National Institute for Occupational Safety and Health, HEW Publications No. (NIOSH) 75-160, April, 1975, Appendix C.

Vickers, R. R. The relationship of defenses and coping to job stress, psychological strain, and coronary heart disease risk factors (doctoral dissertation). University of Michigan, 1979.

Wanous, J. P. and Lawler, E. E., III. Measurement and meaning of job satisfaction. *Journal of Applied Psychology*, 1972, **56**, 95–105.

Zung, W. W. A self rating depression scale. *Archives of General Psychiatry*, 1965, **12**, 63–70.

Appendix A

Occupational Differences

Table A.1   Occupational differences in demography, personality,

	Number of men	Age (years)	Schooling (years)	Duncan SES (percentile)	Income ($)	Length of service
FORKLIFT DRIVER	46	40	11	17	10271	+
ASSEMB MACH-PACED	79	30	12	21	9790	---
ASSEMB M-P RELIEF	27	33	11	21	10140	0
ASSEMB NON-M-PACD	69	39	12	21	11260	–
MACHINE TENDER	34	34	12	22	11548	0
CONTINUOUS FLOW	101	45	12	23	12556	+
COURIER	20	45	11	32	8747	0
TOOL AND DIE	77	49	12	50	12889	+
ELECTRONIC TECH	93	39	13	62	14725	0
POLICEMAN	111	30	14	40	12530	–
SUPVISOR BLUE-COLL	178	42	13	53	14779	0
SUPVISOR WHTE-COLL	42	40	15	78	18494	0
TRAIN DISPATCHER	86	45	13	40	13801	+ +
ATC, LARGE AIRPORT	82	35	13	69	20754	+
ATC, SMALL AIRPORT	43	33	13	69	15764	0
PROGRAMMER	90	33	16	65	14269	–
ACCOUNTANT	92	39	15	78	10802	–
ENGINEER	110	38	17	87	17321	–
SCIENTIST	118	40	18	80	20011	0
PROFESSOR	74	44	19	84	23827	0
ADMIN. PROFESSOR	25	50	19	84	32076	0
ADMINISTRATOR	253	42	16	62	26317	–
PHYSICIAN	104	47	19	92	50813	+ +

Note   Variables described by pluses, zeros, and minuses indicate occupational standings relative to the mean of the random stratified sample:
   + + + at least one standard deviation above the random sample mean
   + + at least two-thirds of a s.d. above the random sample mean
   + at least one-third of a s.d. above the random sample mean

stresses, psychological strains, and health-related behaviors

Type A	Flexibility	Assert Good	Deny Bad	Hrs Worked/Wk (hours)	Hrs Overtime/Wk (hours)	Unwanted Overtime (hours)	Quantitative Workload-E	Combined Qnt Workload	Variance in Workload	Responsibility for Persons-E
0	--	+	0	40.4	3.5	1.2	0	0	0	0
-	0	0	0	41.1	4.0	1.1	0	0	--	-
0	0	0	0	40.5	3.0	0.2	0	0	-	-
0	-	0	0	41.9	5.7	1.9	0	-	-	-
0	-	0	0	42.9	4.3	0.5	0	0	0	0
-	0	0	0	40.8	4.4	0.4	-	--	0	0
0	-	+	0	39.1	1.7	0.5	0	0	0	0
+	-	0	0	46.9	4.3	2.1	0	0	0	-
0	0	0	0	40.2	2.2	0.7	0	0	0	-
0	0	0	0	46.1	6.4	1.5	0	0	+	+
0	0	0	0	47.6	6.8	3.3	0	0	0	+
0	0	0	0	43.7	3.8	2.0	0	0	0	+
0	-	0	0	41.7	2.8	1.6	+	+ +	+	+
0	0	0	0	38.1	0.4	0.5	0	0	+ +	0
0	0	0	0	38.7	0.9	0.0	0	0	+ +	0
-	+	-	0	42.2	3.2	1.2	-	-	0	--
0	0	0	0	40.6	1.9	1.2	0	0	0	--
0	+	0	0	43.3	3.6	1.4	0	0	0	--
0	+ +	0	0	46.6	5.0	1.1	0	0	-	-
0	+ + +	-	-	51.6	3.6	1.5	0	0	0	0
+ +	+ + +	-	-	56.4	11.4	3.6	+ +	+ +	0	+ +
0	0	0	0	48.7	6.4	1.5	0	0	0	+
+	+	0	0	58.4	6.7	3.2	+ +	+ + +	0	+ + +

0 within one-third of a s.d. of the random sample mean
- at least one-third of a s.d. below the random sample mean
-- at least two-thirds of a s.d. below the random sample mean
--- at least one standard deviation below the random sample mean

Table A.1 (continued)   Occupational differences in demography,

	Somatic Complaints	Anxiety	Depression	Irritation	Percent who Smoke (%)	Percent of Smokers Who Have Quit (%)
FORKLIFT DRIVER	0	0	0	0	69	24
ASSEMB MACH-PACED	+ +	+	+	0	60	14
ASSEMB M-P RELIEF	+ +	0	+	0	70	14
ASSEMB NON-M-PACD	+	0	0	0	61	25
MACHINE TENDER	+	+	+	+	62	22
CONTINUOUS FLOW	0	0	–	0	50	38
COURIER	–	0	0	0	47	40
TOOL AND DIE	0	0	0	0	48	42
ELECTRONIC TECH	0	+	0	0	51	30
POLICEMAN	0	0	0	0	50	20
SUPVISOR BLUE-COLL	0	0	0	0	60	30
SUPVISOR WHTE-COLL	0	0	–	0	40	49
TRAIN DISPATCHER	0	0	0	+	59	29
ATC, LARGE AIRPORT	+	0	0	0	59	25
ATC, SMALL AIRPORT	+	0	–	0	65	22
PROGRAMMER	0	0	0	0	31	40
ACCOUNTANT	0	0	0	0	46	35
ENGINEER	0	0	0	0	34	47
SCIENTIST	0	0	0	0	28	41
PROFESSOR	–	0	0	0	34	43
ADMIN. PROFESSOR	0	0	0	+	43	41
ADMINISTRATOR	0	0	0	0	42	37
PHYSICIAN	0	–	0	0	32	46

personality, stresses, psychology strains, and health-related behaviors

Avg Nmbr of Cigarettes Smoked (#)	Avg Number of cups of Coffee (#)	Avg Number of Caffeine Drinks (#)	Recency of Dispensary Visit	Recency of Staffed Dispensary Visit	Obesity
17.0	2.3	4.0	0	0	0
24.6	1.9	4.6	0	+	0
25.7	2.5	4.9	+	+	–
28.0	3.3	5.1	0	0	0
19.1	2.4	3.7	0	0	0
16.4	4.3	5.6	0	–	0
25.9	2.7	4.0	+	+	0
32.2	3.6	4.3	+	0	0
25.1	4.0	5.3	–	0	0
23.9	2.9	4.4	0	+	0
27.2	4.0	5.2	0	0	0
27.2	3.0	4.0	0	0	0
28.3	4.1	5.5	–	–	0
25.5	4.2	5.8	0	0	0
24.9	4.0	5.4	+ +	0	0
21.7	3.1	4.1	0	0	0
23.9	2.9	3.9	–	0	0
24.0	3.1	4.1	0	–	0
17.7	2.5	3.5	–	–	0
27.6	2.9	3.9	–	0	–
33.7	3.5	4.7	–	–	0
28.7	3.4	4.2	0	0	0
27.8	3.0	4.0	0	0	0

Table A.1 (continued)   Occupational differences in demography,

	Job Complexity-E	Concentration	Role Conflict	Role Ambiguity-E	Job Future Ambiguity	Underutilization of Abilities	Inequity of Pay	Income as % of Deserved Income
FORKLIFT DRIVER	--	--	+	0	+	+ +	0	0
ASSEMB MACH-PACED	---	---	0	0	+ +	+ + +	-	0
ASSEMB M-P RELIEF	--	--	+	0	0	+ +	-	--
ASSEMB NON-M-PACD	--	--	0	0	+	+ +	0	0
MACHINE TENDER	--	0	0	0	+ +	+ +	0	0
CONTINUOUS FLOW	0	0	0	0	0	+	-	0
COURIER	--	0	0	-	0	+	0	0
TOOL AND DIE	0	0	0	0	+	-	+	-
ELECTRONIC TECH	0	0	0	0	0	-	+	0
POLICEMAN	+ +	0	0	0	-	-	0	-
SUPVISOR BLUE-COLL	+	0	0	0	0	0	0	0
SUPVISOR WHTE-COLL	+	0	0	0	0	0	0	0
TRAIN DISPATCHER	+	+ +	+	-	0	0	0	0
ATC, LARGE AIRPORT	0	+ +	0	-	-	0	0	-
ATC, SMALL AIRPORT	0	+ +	-	--	--	0	0	0
PROGRAMMER	0	0	0	+	0	0	0	0
ACCOUNTANT	0	0	0	0	0	0	+	0
ENGINEER	0	0	0	+	0	0	0	0
SCIENTIST	0	0	0	+	0	-	0	0
PROFESSOR	+	0	-	0	0	---	0	0
ADMIN. PROFESSOR	+ +	0	0	0	-	--	0	0
ADMINISTRATOR	+ +	0	0	0	0	-	0	0
PHYSICIAN	+ + +	+ +	-	-	--	---	0	0

personality, stresses, psychological strains, and health-related behaviors

Participation	Social Support from Supervisor	Social Support from Others at Work	Social Support from Home	Workload Fit	Responsibility for Persons Poor Fit	Job Complexity Poor Fit	Role Ambiguity Poor Fit	Job Dissatisfaction	Boredom	Workload Dissatisfaction
−	−	−	0	0	+	+	+	+	+ +	+
−	−	−	0	0	+	+ + +	0	+ +	+ + +	+ +
−	−	− −	−	0	0	0	0	+	+ +	+
−	−	0	−	0	+	+ +	+	+	+ +	0
− −	−	− −	0	+	0	+ +	+ +	+ +	+ +	+ +
0	+	+	0	−	0	0	0	+ +	0	0
−	0	0	0	0	+ +	0	+	−	0	−
0	0	0	−	0	+	0	0	0	0	0
0	0	0	−	0	+	0	0	0	0	0
0	0	0	0	0	−	−	0	−	−	0
+	0	0	0	0	−	−	0	−	−	0
0	0	0	0	+	0	0	0	0	−	0
0	0	0	0	−	0	0	0	0	−	−
0	0	+	0	−	0	0	−	−	−	0
0	0	0	0	−	0	0	0	0	0	0
0	0	0	0	0	+	0	0	0	0	0
0	0	0	0	0	0	0	0	0	−	0
0	+	0	0	+	−	−	0	−	−	0
+ +	+	+ +	+	+ +	0	−	0	−	−	+
+	0	0	0	0	−	−	0	0	−	0
0	+	+	0	+	−	−	0	−	−	+

Description of Five Measures not Included in Appendix E of Caplan *et al.* (1980)

Typical Education Needed-E (Education-E)

Item content and format

What level of formal education do you feel is needed by a person in a job such as yours? (Check one box) (Ignore numbers in brackets []).

None [1]
Grades 1–4 [2]
Grades 5–6 [3]
Grades 7–8 [4]
Grades 9–11 [5]
Grade 12 (completed high school) [6]
Completed high school plus other non-college training
(technical or trade school) [7]
Some college [8]
Completed college with bachelor's degree [9]
Completed college with advanced or professional degree
(MA, MS, PhD, MD, etc.) [10]

Typical Service Needed-E (Service-E)

Item content and format

How long do you feel a person needs to work in your particular job to be fully trained? (Check one box)

Less than one month [1]	Between 1 and 5 years [4]
1 to six months [2]	Between 6 and 10 years [5]
7 months to a year [3]	More than 10 years [6]

Number Supervised

Item content and format

Do you supervise anyone? (Check one box)

No	Yes, I directly supervise _____ people
[1]	[2] (fill in number)

Job Insecurity

Item content and format

How likely is it that in the next few years your job will be replaced by computers or other machines or that it may be eliminated or given to someone else? (Circle one number)

1	2	3	4
Not at all likely	A little likely	Somewhat likely	Very likely

Inequity of Utilization

Item content and format

Compared to others in your organization, do you get a fair share of the opportunities to use your best skills, knowledge, abilities, and your own ideas? (Circle one number)

1	2	3	4	5
Very much less than I ought to get	Somewhat less than I ought to get	A little less than I ought to get	About the same as I ought to get	More than I ought to get

Correlations between Strains and the E, P, Fit, Deficiency, Excess, and Poor Fit Measures on Eight Dimensions

Table C.1 Correlations between strains and the job complexity variables

| | | | | Job Complexity | | | |
Strains	E	P	Fit	De-ficiency	Excess	Poor Fit	$n \geq$
Job Dissatisfaction	−.31**	−.30**	−.03	−.19**	.19**	.47**	309
Workload Dissatisfaction	−.04	−.17**	.15**	−.03	.32**	.36**	308
Boredom	−.51**	−.34**	−.26**	−.38**	.02	.51	308
Depression	−.09	−.12*	.03	−.09	.17**	.22**	309
Anxiety	.00	−.05	.06	−.08	.21**	.21**	309
Irritation	.00	−.08	.09	−.02	.19**	.15**	309
Somatic Complaints	−.11	−.02	−.11	−.19**	.04	.16**	308
Number of Cigarettes Smoked	−.07	.01	−.11	−.20*	.05	.18*	146
Coffee and Tea	.06	.06	.00	.00	.00	.09	261
Obesity	.02	.07	−.07	−.04	−.08	.04	304
Heart Rate (A)	.09	.12*	−.03	.02	−.09	−.11*	333
Systolic Blood Pressure (A)	−.09	−.10	.01	.01	−.00	−.01	328
Diastolic Blood Pressure (A)	−.11*	−.10	−.02	−.02	−.02	.02	371
Cholesterol (A)	.01	.02	−.02	−.02	−.02	−.01	342
T3	.08	.12*	−.05	−.11*	.05	.04	327
T4I	.09	.09	.01	.00	.02	−.02	328
Serum Uric Acid (A)	−.01	.01	−.02	.02	−.07	−.01	328
Cortisol (A)	−.02	−.02	.01	.03	−.02	−.07	340

Note The physiological sample ($n = 390$) was used to determine correlations involving the eight physiological measures. The random stratified sample ($n = 318$) was used to determine the other correlations.

E = reported environmental level; P = reported preferred level; Fit = E–P; Deficiency = E–P for E–P $\leq$ 0 and Deficiency = 0 for E–P $>$ 0; Excess = E–P for E–P $\geq$ 0 and Excess = 0 for E–P $<$ 0; and Poor Fit = $|$E–P$|$.

$*p < .05; **p < .01$

Table C.2   Correlations between strains and the role ambiguity variables

Strains	E	P	Fit	De-ficiency	Excess	Poor Fit	$n \geq$
Job Dissatisfaction	.17**	.04	.07	−.01	.16**	.19**	307
Workload Dissatisfaction	.16**	.09	.03	−.02	.10	.13*	306
Boredom	.10	.03	.03	−.04	.12*	.17**	306
Depression	.19**	.07	.06	.01	.10	.12*	307
Anxiety	.17**	.02	.09	.07	.08	.01	307
Irritation	.09	−.06	.12	.08	.14*	.06	307
Somatic Complaints	.11	.05	.03	.03	.02	.02	306
Number of Cigarettes Smoked	−.07	−.11	.06	.03	.10	.02	149
Coffee and Tea	.01	−.00	.01	.03	−.02	−.06	158
Obesity	.07	.02	.04	−.00	.08	.02	302
Heart Rate (A)	−.04	.02	−.04	−.01	−.06	−.03	338
Systolic Blood Pressure (A)	−.14*	−.09	−.01	.08	−.11*	−.14*	333
Diastolic Blood Pressure (A)	−.12*	.02	−.10*	−.03	−.15**	−.04	376
Cholesterol (A)	−.00	.05	−.06	−.03	−.07	−.03	345
T3	.09	.02	.05	−.00	.09	.08	329
T4I	.16**	.07	.06	.01	.10	.07	330
Serum Uric Acid (A)	−.04	−.09	.04	.08	−.03	−.09	331
Cortisol (A)	.01	.06	−.04	−.02	−.03	.03	343

Note   The physiological sample ($n = 390$) was used to determine correlations involving the eight physiological measures. The random stratified sample ($n = 318$) was used to determine the other correlations.

E = reported environmental level; P = reported preferred level; Fit = E−P; Deficiency = E−P for E−P $\leq$ 0 and Deficiency = 0 for E−P > 0; Excess = E−P for E−P $\geq$ 0 and Excess = 0 for E−P < 0; and Poor Fit = $|E-P|$.

$*p < .05; **p < .01$

Table C.3   Correlations between strains and the responsibility for persons variables

Strains	E	P	Fit	De-ficiency	Excess	Poor Fit	$n \geq$
				Responsibility for Persons			
Job Dissatisfaction	−.26**	−.15**	−.14*	−.18**	.02	.23**	309
Workload Dissatisfaction	−.05	−.15**	.06	−.00	.17**	.07	388
Boredom	−.30**	−.07	−.24**	−.29**	−.03	.32**	308
Depression	−.13*	−.16**	.01	−.02	.08	.05	310
Anxiety	−.04	−.11	.05	.01	.12*	.05	310
Irritation	−.02	−.09	.06	.04	.06	.00	310
Somatic Complaints	−.13*	−.03	−.09	−.09	−.05	.07	308
Number of Cigarettes Smoked	−.15	−.03	−.11	−.13	−.01	.14	150
Coffee and Tea	−.04	−.05	.01	.01	.00	.03	262
Obesity	−.05	.03	−.09	−.07	−.09	−.01	305
Heart Rate (A)	.04	.03	.01	.03	−.04	−.05	336
Systolic Blood Pressure (A)	.00	.04	−.03	−.05	.06	.09	331
Diastolic Blood Pressure (A)	−.01	.05	−.05	−.06	.02	.07	374
Cholesterol (A)	.09	.01	.09	.08	.05	−.03	343
T3	−.07	−.04	−.04	−.02	−.08	−.04	327
T4I	.03	−.01	.04	.05	−.00	−.07	328
Serum Uric Acid (A)	.10	.10	.03	.01	.05	.04	329
Cortisol (A)	.06	−.02	.07	.05	.09	−.08	341

Note   The physiological sample (n = 390) was used to determine correlations involving the eight physiological measures. The random stratified sample (n = 318) was used to determine the other correlations.

E = reported environmental level; P = reported preferred level; Fit = E–P; Deficiency = E–P for E–P $\leq$ 0 and Deficiency = 0 for E–P > 0; Excess = E–P for E–P $\geq$ 0 and Excess = 0 for E–P < 0; and Poor Fit = |E–P|.

p < .05; **p* < .01

Table C.4   Correlations between strains and the workload variables

Strains	E	P	Fit	Workload De-ficiency	Excess	Poor Fit	$n \geq$
Job Dissatisfaction	.01	−.26**	.19**	.07	.21**	.22**	300
Workload Dissatisfaction	.33**	−.34**	.52**	.26**	.54**	.36**	299
Boredom	−.19**	−.28**	.05	.02	.06	.12*	299
Depression	.05	−.29**	.27**	.16**	.26**	.24**	301
Anxiety	.05	−.11	.13*	.05	.15**	.09	301
Irritation	.21**	−.15**	.27**	.14*	.29**	.17**	301
Somatic Complaints	.00	−.06	.05	.01	.06	.06	299
Number of Cigarettes Smoked	−.10	−.08	−.05	−.01	−.07	.08	148
Coffee and Tea	.01	.05	−.03	−.01	−.03	.03	253
Obesity	.04	.13*	−.06	−.07	−.04	−.03	296
Heart Rate (A)	.05	−.05	.06	.07	.02	−.07	329
Systolic Blood Pressure (A)	.07	−.07	.11*	.07	.10	.03	324
Diastolic Blood Pressure (A)	.00	.01	−.01	−.03	.01	.03	366
Cholesterol (A)	.07	.04	.01	−.03	.04	−.01	336
T3	.06	.08	.00	.03	−.02	−.11*	320
T4I	.11*	.02	.07	.07	.06	−.01	321
Serum Uric Acid (A)	.04	−.04	.03	−.02	.06	.07	322
Cortisol (A)	.01	−.05	.04	.06	.01	−.02	334

Note   The physiological sample ($n = 390$) was used to determine correlations involving the eight physiological measures. The random stratified sample ($n = 318$) was used to determine the other correlations.

E = reported environmental level; P = reported preferred level; Fit = E–P; Deficiency = E–P for E–P $\leq$ 0 and Deficiency = 0 for E–P > 0; Excess = E–P for E–P $\geq$ 0 and Excess = 0 for E–P < 0; and Poor Fit = $|E-P|$.

*$p < .05$; **$p < .01$

Table C.5   Correlations between strains and the income variables

Strains	Income			$n \geq$
	E	P	Fit	
Job Dissatisfaction	−.25**	−.23**	−.11	279
Workload Dissatisfaction	−.05	.06	−.23**	279
Boredom	−.32**	−.34*	−.10	279
Depression	−.14*	−.11	−.12*	280
Anxiety	−.05	−.05	−.07	280
Irritation	−.04	−.02	−.05	280
Somatic Complaints	−.16**	−.12*	−.17**	279
Number of Cigarettes Smoked	−.09	−.09	.11	138
Coffee and Tea	−.05	−.01	.03	239
Obesity	.03	.04	−.08	277
Heart Rate (A)	.07	.06	.02	306
Systolic Blood Pressure (A)	−.04	−.02	.01	300
Diastolic Blood Pressure (A)	−.05	−.00	−.04	339
Cholesterol (A)	.04	.07	−.01	312
T3	.08	.08	.02	301
T4I	−.00	−.03	.04	302
Serum Uric Acid (A)	−.07	.01	−.14*	297
Cortisol (A)	−.01	−.08	.14*	311

Note   The physiological sample ($n = 390$) was used to determine correlations involving the eight physiological measures. The random stratified sample ($n = 318$) was used to determine the other correlations.

E = reported environmental level; P = reported preferred level; and Fit = (E–P)/P.

$*p < .05; **p < .01.$

Table C.6   Correlations between strains and the overtime variables

Strains	Overtime E	P	Fit	$n \geq$
Job Dissatisfaction	.09	.01	.19**	304
Workload Dissatisfaction	.20**	.05	.38**	303
Boredom	.01	−.02	.06	303
Depression	.14*	.09	.10	303
Anxiety	.10	.01	.18*	303
Irritation	.03	−.05	.04	303
Somatic Complaints	.06	.01	.13*	302
Number of Cigarettes Smoked	−.01	−.03	.04	150
Coffee and Tea	.07	.11	−.07	255
Obesity	.05	.01	.09	298
Heart Rate (A)	.02	.04	−.03	338
Systolic Blood Pressure (A)	−.04	−.05	.02	333
Diastolic Blood Pressure (A)	−.03	−.06	.03	376
Cholesterol (A)	.02	.02	.02	345
T3	.13*	.11*	.09	329
T4I	.02	.05	−.01	330
Serum Uric Acid (A)	.03	.02	.02	331
Cortisol (A)	.16**	.14**	.12*	343

Note   The physiological sample ($n = 390$) was used to determine correlations involving the eight physiological measures. The random stratified sample ($n = 318$) was used to determine the other correlations.

E = reported environmental level; P = reported preferred level; and Fit = E-P.

*$p < .05$; **$p < .01$

Table C.7   Correlations between strains and the length of service variables

Strains	E	P	Fit	Tenure De-ficiency	Excess	Poor Fit	$n \geq$
Job Dissatisfaction	−.22**	−.09	−.14*	−.16**	−.07	.07	311
Workload Dissatisfaction	−.10	.02	−.12*	−.10	−.09	.00	310
Boredom	−.50**	−.22**	−.28**	−.36**	−.11	.19**	310
Depression	−.09	.04	−.11	−.10	−.09	.00	312
Anxiety	−.05	−.00	−.00	−.03	−.02	.04	312
Irritation	−.05	.02	−.07	−.04	−.07	−.03	312
Somatic Complaints	−.10	−.07	−.01	−.04	.02	.04	310
Number of Cigarettes Smoked	−.10	.01	−.10	−.11	−.05	.07	152
Coffee and Tea	.03	−.02	.03	.03	.02	−.01	263
Obesity	−.00	.09	−.09	−.08	−.06	.01	307
Heart Rate (A)	.01	.10	−.08	−.04	−.09	−.07	338
Systolic Blood Pressure (A)	−.04	.05	−.06	−.08	−.04	.00	332
Diastolic Blood Pressure (A)	−.06	.00	−.01	−.06	.05	.08	375
Cholesterol (A)	.04	−.02	.09	.02	.10	.09	346
T3	−.01	−.11*	.08	.05	.07	.05	329
T4I	−.02	−.12*	.11*	.06	.10	.07	330
Serum Uric Acid (A)	−.02	.16**	−.14*	−.13*	−.11*	−.04	331
Cortisol (A)	−.09	−.01	−.04	−.07	−.01	.03	343

Note   The physiological sample ($n = 390$) was used to determine correlations involving the eight physiological measures. The random stratified sample ($n = 318$) was used to determine the other correlations.

E = reported level necessary to perform; P = reported level of individual; Fit = (E-P)/P; Deficiency = (E-P)/P for (E-P/P ≤ 0 and Deficiency = 0 for (E-P)/P > 0; Excess = (E-P)/P for (E-P)/P ≥ 0 and Excess = 0 for (E-P)/P < 0; and Poor Fit = |(E-P)/P|.

*$p < .05$; **$p < .01$

Table C.8 Correlations between strains and the education variables

Strains	E	P	Fit	De-ficiency	Excess	Poor Fit	$n \geq$
Job Dissatisfaction	−.33**	−.20**	−.27**	−.30**	−.01	.29**	315
Workload Dissatisfaction	−.15*	−.08	−.14*	−.17**	.04	.19**	315
Boredom	−.54**	−.31**	−.47**	−.51**	−.05	.48**	315
Depression	−.16**	−.08	−.15**	−.19**	.07	.22**	315
Anxiety	−.04	−.03	−.03	−.04	.02	.05	315
Irritation	−.06	−.05	−.03	−.05	.03	.06	315
Somatic Complaints	−.21**	−.21**	−.11	−.12*	−.01	.11	315
Number of Cigarettes Smoked	−.28**	−.26**	−.09	−.15	.11	.20*	153
Coffee and Tea	.03	−.04	.07	.05	.09	−.00	267
Obesity	.02	−.07	.10	.08	.10	−.04	311
Heart Rate (A)	.07	.01	.07	.07	.03	−.05	344
Systolic Blood Pressure (A)	−.17**	−.21**	−.01	.00	−.03	−.02	339
Diastolic Blood Pressure (A)	−.16**	−.19**	−.00	−.01	.02	.02	382
Cholesterol (A)	.04	.01	.06	.06	.01	−.05	351
T3	.21**	.28**	−.03	.01	−.10	−.06	335
T4I	.07	.10	−.03	−.06	.04	.07	336
Serum Uric Acid (A)	−.09	−.12*	.03	.03	.01	−.03	337
Cortisol (A)	.00	.01	.03	.00	.07	.03	349

Note The physiological sample (n = 390) was used to determine correlations involving the eight physiological measures. The random stratified sample (n = 318) was used to determine the other correlations.

E = reported level necessary to perform; P = reported level of individual; Fit = (E−P)/P; Deficiency = (E−P)/P for (E−P)/P ≤ 0 and Deficiency = 0 for (E−P)/P > 0; Excess = (E−P)/P for (E−P)/P ≥ 0 and Excess = 0 for (E−P)/P < 0; and Poor Fit = |(E−P)/P|.

*p < .05; **p < .01

Appendix D

Independent Predictors of Strains

Subjective environment variables, personality variables, and P–E fit variables checked for possible independent relationships with each strain include: Job Complexity-E, Role Ambiguity-E, Responsibility for Persons-E, Workload-E, Income-E, Overtime-E, Tenure-E, Education-E, Hours Worked per Week, Workload-Quinn, Workload Variance, Concentration, Role Conflict, Job Future Ambiguity, Job Insecurity, Underutilization of Abilities, Inequity of Pay, Participation, Social Support from Supervisor, Social Support from Others at Work, Social Support from Wife, Friends, and Relatives, Number Supervised, Sales Type A Personality, Flexibility, Assert Good Self, Deny Bad Self, Job Complexity-P, Role Ambiguity, Responsibility for Persons-P, Workload-P, Income-P, Overtime-P, Tenure-P, Education-P, Job Complexity Fit, Job Complexity Deficiency, Job Complexity Excess, Job Complexity Poor Fit, Role Ambiguity Deficiency, Role Ambiguity Excess, Role Ambiguity Poor Fit, Responsibility for Persons Fit, Responsibility for Persons Deficiency, Responsibility for Persons Excess, Responsibility for Persons Poor Fit, Workload Poor Fit, Workload Deficiency, Workload Excess, Workload Poor Fit, Income Fit, Overtime Fit, Service Fit, Service Deficiency, Service Excess, Service Poor Fit, Education Fit, Education Deficiency, Education Excess, and Education Poor Fit.

Table D.1   Partial correlations of Job Dissatisfaction with predictor variables

Predictor variables	Partial correlation in original random stratified sample	Partial correlation replication in second random stratified sample
Job Complexity Poor Fit	.28***	.30***
Other Work Support	-.17**	-.04
Workload Fit	.16**	.04
Overtime Fit	.14*	.08
Inequity of Utilization	.14*	.03
Future Ambiguity	.14*	.30***
Underutilization	.13*	.21***
Participation	-.13*	.03
Multiple correlation	.64***	.61***

Predictor variables	Partial correlation in second random stratified sample	Partial correlation replication in original random stratified sample
Job Complexity Poor Fit	.34***	.34***
Future Ambiguity	.31***	.22***
Underutilization	.28***	.23***
Service-E	.19***	.10
Responsibility for Persons-P	-.16**	-.12*
Flexibility	-.15*	-.10
Service Poor Fit	.12*	.00
Multiple correlation	.65***	.59***

Note   The partial correlation between a predictor and Job Complexity represents the correlation between them, holding constant the other predictors in the same list. For the original random stratified sample analysis $n = 277$. For the second random stratified sample analysis $n = 300$.
  $*p < .05; **p < .01; ***p < .001$

Table D.2   Partial correlations of Workload Dissatisfaction with predictor variables

Predictor variables	Partial correlation in original random stratified sample	Partial correlation replication in second random stratified sample
Workload Excess	.41***	.39***
Job Complexity Poor Fit	.30***	.17**
Overtime Fit	.29***	.21***
Workload-Quinn	.26***	.21***
Other Work Support	−.20***	−.10
Concentration	−.17	−.09
Multiple correlation	.70***	.67***

Predictor variables	Partial correlation in second random stratified sample	Partial correlation replication in original random stratified sample
Workload Excess	.41***	.38***
Overtime Fit	.24***	.26***
Workload-Quinn	.21***	.25***
Deny Bad Self	−.18**	−.19**
Inequity of Utilization	.18**	.06
Job Future Ambiguity	.17**	.13*
Job Complexity-P	−.14*	−.15**
Multiple correlation	−.70***	.65***

Note   The partial correlation between a predictor and Workload Dissatisfaction represents the correlation between them, holding constant the other predictors in the same list. For the original random stratified sample analysis $n = 280$. For the second random stratified sample analysis $n = 286$.

$*p < .05; **p < .01; ***p < .001$

Table D.3   Partial correlation of Boredom with predictor variables

Predictor variables	Partial correlation in original random stratified sample	Partial correlation replication in second random stratified sample
Job Complexity Poor Fit	.29***	.26***
Underutilization	.21***	.31***
Concentration	−.18**	−.21***
Education Poor Fit	.18**	.27***
Deny Bad Self	.15**	−.13*
Job Complexity-E	−.13*	−.19***
Service-E	−.12*	.00
Multiple correlation	.73***	.73***

Predictor variables	Partial correlation in second random stratified sample	Partial correlation replication in original random stratified sample
Education Deficiency	−.34***	−.25***
Job Complexity Poor Fit	.33***	.35***
Underutilization	.32***	.28***
Job Complexity Excess	−.21***	−.15*
Workload-Quinn	−.19**	−.02
Job Future Ambiguity	.19**	.09
Service Poor Fit	−.16**	.08
Overtime-P	−.14*	−.01
Supervisor Support	−.13*	−.06
Multiple correlation	.76***	.69***

Note The partial correlation between a predictor and Boredom represents the correlation between them, holding constant the other predictors in the same list. For the original random stratified sample analysis $n = 297$. For the second random stratified sample analysis $n = 274$.
   $*p < .05; **p < .01; ***p < .001$

Table D.4　Partial correlations of Depression with predictor variables

Predictor variables	Partial correlation in original random stratified sample	Partial correlation replication in second random stratified sample
Other Work Support	−.27***	−.36***
Workload Fit	.26***	.13*
Assert Good Self	−.23***	−.11
Job Future Ambiguity	.18**	.25***
Multiple correlation	.49***	.50***

Predictor variables	Partial correlation in second random stratified sample	Partial correlation replication in original random stratified sample
Other Work Support	−.35***	−.28***
Role Conflict	.27***	.17**
Deny Bad Self	−.21***	−.20***
Job Future Ambiguity	.19**	.12*
Schooling-P	.18**	.06
Income Fit	−.13*	−.06
Multiple correlation	.59***	.46***

Note The partial correlation between a predictor and Depression represents the correlation between them, holding constant the other predictors in the same list. For the original random stratified sample analysis $n = 298$. For the second random stratified sample analysis $n = 283$.
　$*p < .05; **p < .01; ***p < .001$

Table D.5 Partial correlations of Anxiety with predictor variables

Predictor variables	Partial correlation in original random stratified sample	Partial correlation replication in second random stratified sample
Deny Bad Self	−.28***	−.20***
Role Conflict	.20***	.17**
Other Work Support	−.19***	−.23***
Workload-Quinn	.17**	.15**
Concentration	−.15**	.05
Responsibility for Persons Excess	.15**	−.02
Multiple correlation	.47***	.42***

Predictor variables	Partial correlation in second random stratified sample	Partial correlation replication in original random stratified sample
Workload-Quinn	.22***	.17**
Income-Fit	−.20***	.00
Deny Bad Self	−.20***	−.33***
Home Support	−.16**	−.02
Role Ambiguity Fit	.16**	.05
Multiple correlation	.45***	.37***

Note The partial correlation between a predictor and Anxiety represents the correlation between them, holding constant the other predictors in the same list. For the original random stratified sample analysis $n = 304$. For the second random stratified sample analysis $n = 275$.
  $*p < .05; **p < .01; ***p < .001$

Table D.6   Partial correlations of Irritation with predictor variables

Predictor variables	Partial correlation in original random stratified sample	Partial correlation replication in second random stratified sample
Deny Bad Self	−.31***	−.30***
Role Conflict	.28***	.29***
Workload Excess	.21***	.12*
Other Work Support	−.13*	−.17**
Multiple correlation	.51***	.51***

Predictor variables	Partial correlation in section random stratified sample	Partial correlation replication in original random stratified sample
Role Conflict	.31***	.33***
Deny Bad Self	−.26***	−.34***
Underutilization	.18**	−.06
Other Support	−.17**	−.14*
Type A Personality	.17**	.09
Responsibility for Persons Excess	.16**	.05
Service Deficiency	.15**[a]	−.01
Multiple correlation	.54***	.51***

Note   The partial correlation between a predictor and Irritation represents the correlation between them, holding constant the other predictors in the same list. For the original random stratified sample analysis $n = 298$. For the second random stratified sample analysis $n = 299$.

[a]This relationship is opposite to the prediction that the larger the discrepancy the higher the strain.

*$p < .05$; **$p < .01$; ***$p < .001$

Table D.7 Partial correlations of Somatic Complaints with predictor variables

Predictor variables	Partial correlation in original random stratified sample	Partial correlation replication in second random stratified sample
Role Conflict	.22***	.15**
Deny Bad Self	−.22***	−.19***
Education-P	−.20***	−.21***
Job Complexity Deficiency	−.18***	−.11
Workload Variance	.14*	−.02
Flexibility	.12*	−.03
Multiple correlation	.44***	.36***

Predictor variables	Partial correlation in second random stratified sample	Partial correlation replication in original random stratified sample
Deny Bad Self	−.23***	−.21***
Income Fit	−.19**	−.13*
Responsibility for Persons-E	.17**	−.09
Job Complexity-E	−.16**	−.02
Schooling-P	−.15**	−.17**
Workload Excess	.15*	.06
Multiple correlation	.40***	.35***

Note The partial correlation between a predictor and Somatic Complaints represents the correlation between them, holding constant the other predictors in the same list. For the original random stratified sample analysis $n = 304$. For the second random stratified sample analysis $n = 265$.

$*p < .05; **p < .01; ***p < .001$

Table D.8 Partial correlations of Number of Cigarettes Smoked with predictor variables

Predictor variables	Partial correlation in original random stratified sample	Partial correlation replication in second random stratified sample
Education-E	−.32***	−.17*
Assert Good Self	−.20*	−.24**
Education Excess	.19*	.02
Job Complexity-E	.17*	.01
Workload Variance	−.17*	.11
Multiple correlation	.41***	.30*

Predictor variable	Correlation in second random stratified sample	Correlation replication in original random stratified sample
Assert Good Self	−.19*	−.15

Note The partial correlation between a predictor and Number of Cigarettes Smoked represents the correlation between them, holding constant the other predictors in the same list. For the original random stratified sample analysis $n = 144$. For the second random stratified sample analysis $n = 157$.

*p < .05; **p < .01; ***p < .001

Table D.9   Partial correlations of Coffee and Tea with predictor variables

Predictor variables	Partial correlation in original random stratified sample	Partial correlation replication in second random stratified sample
(No significant predictors)		

Predictor variables	Partial correlation in second random stratified sample	Partial correlation replication in original random stratified sample
Role Conflict	.25***	.12*
Role Ambiguity-E	−.21***	.00
Concentration	.19**	.00
Type A Personality	−.16**	.01
Workload-P	.14*	.05
Service Poor Fit	−.14*[a]	−.01
Multiple correlation	.37***	.14

Note   The partial correlation between a predictor and Coffee and Tea represents the correlation between them, holding constant the other predictors in the same list. For the original random stratified sample analysis $n = 311$. For the second random stratified sample analysis $n = 249$.
[a]This relationship is opposite to the prediction that the larger the poor fit, the higher the strain.
*$p < .05$; **$p < .01$; ***$p < .001$

Table D.10   Partial correlations of Obesity with predictor variables

Predictor variables	Correlation in original random stratified sample	Correlation replication in second random stratified sample
Workload-P	.13*	−.02

Predictor variables	Partial correlation in second random stratified sample	Partial correlation replication in original random stratified sample
Service-P	.17**	.10
Education-P	−.14*	−.07
Responsibility for Persons-P	.13*	.02
Multiple correlation	.27***	.13

Note   The partial correlation between a predictor and Obesity represents the correlation between them, holding constant the other predictors in the same list. For the original random stratified sample analysis $n = 309$. For the second random stratified sample analysis $n = 304$.
*$p < .05$; **$p < .01$; ***$p < .001$

152

Table D.11   Correlation of Heart Rate (A) with predictor
variables

Predictor variable	Correlation
Job Complexity Poor Fit	−.11*[a]

Note   For this physiological sample analysis $n = 333$.
[a]This relationship is opposite to the prediction that the larger the poor fit, the higher the strain.
*$p < .05$

Table D.12   Partial correlations of Systolic Blood Pressure
(A) with predictor variables

Predictor variables	Partial correlation
Number Supervised	−.18***
Workload-E	.13*
Other Work Support	.12*
Role Ambiguity Poor Fit	−.12*[a]
Multiple correlation	.26***

Note   The partial correlation between a predictor and Systolic Blood Pressure represents the correlation between them, holding constant the other predictors in the same list. For this physiological sample analysis $n = 329$.
[a]This relationship is opposite to the prediction that the larger the poor fit, the higher the strain.
*$p < .05$; **$p < .01$; ***$p < .001$

153

Table D.13 Partial correlations of Diastolic Blood Pressure (A) with predictor variables

Predictor variables	Partial correlation
Education-E	−.18***
Role Ambiguity Excess	−.17***[a]
Multiple correlation	.23***

Note The partial correlation between a predictor and Diastolic Blood Pressure represents the correlation between them, holding constant the other predictors in the same list. For this physiological sample analysis $n = 376$.

[a]This relationship is opposite the prediction that the larger the excess, the higher the strain.

***$p < .001$

Table D.14 Correlation of Cholesterol (A) with predictor variables

Predictor variable	Correlation
Type A Personality	.12*

Note For this physiological sample analysis $n = 351$.

*$p < .05$

Table D.15 Partial correlations of T3 with predictor variables

Predictor variables	Partial correlation
Education-P	.30***
Job Complexity Poor Fit	.11*
Multiple correlation	.30***

Note The partial correlation between a predictor and T3 represents the correlation between them, holding constant the other predictors in the same list. For this physiological sample analysis $n = 327$.

*$p < .05$; ***$p < .001$

Table D.16   Partial correlations of T4I with predictor variables

Predictor variables	Partial correlation
Job Insecurity	.17**
Service Fit	.14**
Multiple correlation	.20***

Note   The partial correlation between a predictor and T4I represents the correlation between them, holding constant the other predictors in the same list. For this physiological sample analysis $n = 330$.
$**p < .01; ***p < .001$

Table D.17   Partial correlations of Serum Uric Acid (A) with predictor variables

Predictor variables	Partial correlation
Service-P	.16**
Job Insecurity	−.12*
Multiple correlation	.20***

Note   The partial correlation between a predictor and Serum Uric Acid represents the correlation between them, holding constant the other predictors in the same list. For this physiological sample analysis $n = 334$.
$*p < .05; **p < .01; ***p < .001$

Table D.18   Partial correlations of Cortisol (A) with predictor variables

Predictor variables	Partial correlation
Overtime-E	.17***
Role Conflict	−.14**
Concentration	−.14**
Multiple correlation	.25***

Note   The partial correlation between a predictor and Cortisol represents the correlation between them, holding constant the other predictors in the same list. For this physiological sample analysis $n = 343$.
$**p < .01; ***p < .001$

Author Index

156

Subject Index

158

Hours worked per week
measurement of, 12, 14, 18

Illness, 3, 5–7, 23, 27, 69–72, 92, *see also* Cardiovascular disease, Gastro-intestinal problems, Peptic ulcer, Psychosomatic illness, Respiratory infection
Income, *see also* Inequity of pay
measurement of, 12, 14–16, 31–39
and strain, 33, 43, 44, 45, 47
Inequity of pay, *see also* Income
measurement of, 15, 18
Inequity of utilization
measurement of, 15
Intervening variable, 57, 59–72, 94–96, 98, 100, 103
Irritation, 22, 30, 31, 40–44, 50–51, 54–55, 60, 65–68, 70, 73, 107

Job complexity, 7, 75, 76, 92, 95
measurement of, 14–18, 32–39
and strain, 32, 40, 42, 45, 50, 54–55, 59, 60, 63, 64, 66–67, 71, 79–84, 87
Job demands, *see also* Person–environment fit, stress
qualitative versus quantitative, 106
Job future ambiguity
measurement of, 14, 18
and strain, 54, 60, 63, 66–67, 92, 93, 107
Job insecurity
measurement of, 14
and strain, 55
Job-related strain, 1–7, 68, 98, 100, 101, *see also* Job satisfaction–dissatisfaction, Dissatisfaction with workload, Boredom
Job satisfaction–dissatisfaction, 1, 3, 5–7, 13, 21, 30–32, 40–44, 50–51, 54–55, 60, 63–68, 70–71, 73, 75, 87, 90, 91, 99, 107

Length of service
measurement of, 12, 14, 16, 19–21, 32–39
and strain, 32, 43–45, 49, 51, 55, 58

Machine-paced assemblers, 11, 13, 83–84, 87, 91–94, 103
Mediating variable, *see* Intervening variable
Mental health, 1, 2, 7
Morbidity, 6, 7, *see also* Illness

Mortality, 6, 7, *see also* Illness

Need for social approval, 19
Noise, 9

Obesity, 17, 26, 42–43, 50–51, 54
Objective versus subjective, 2–4, 6–8, 102, 109
Occupation, 1, 6, 7, 9–11, 14, 93–100, 102, *see also* Administrators, Air traffic controllers, Blue collar workers, Delivery service couriers, Forklift drivers, Machine-paced assemblers, Policemen, Scientists, Train dispatchers, University professors, White collar workers
and relationships with strain, 83–101, 104
Overtime
measurement of, 14–18, 20, 32–39
and strain, 32, 41, 43–45, 47, 54–55, 60, 63, 67

Participation, 92, 93, 106, 109
measurement of, 3, 15, 18, 74, 75, 109
and strain, 57, 60, 63–68, 85, 87, 106, 109
Pay, *see* Income
inequity of, *see* Inequity of pay
Peptic ulcer, 9, 17, 101
Perception, *see* Accessibility of the self, Contact with reality
Personality, 9, 18, 19, 22, *see also* Assert good self, Deny bad self, Flexibility, Need for social approval, self-esteem, Self-identity, Type A
Person–environment fit–misfit, 1–6, 27–58, 94, 95, 100, 106, 109, *see also* Application of findings, Education, Income, Job Complexity, Length of service, Overtime, Responsibility for persons, Role ambiguity, Work load
and explaining additional variance in strains, 39, 46–52, 107
and commensurate dimensions, 4, 19, 64, 89, 108
and correlations with strains, 3, 9–46, 74–86
and independent predictors of strains, 52–58
and interactions, 82–84, 103
and measurement issues, 21, 29–31, 33–39
measures of, 19–21, 33–38